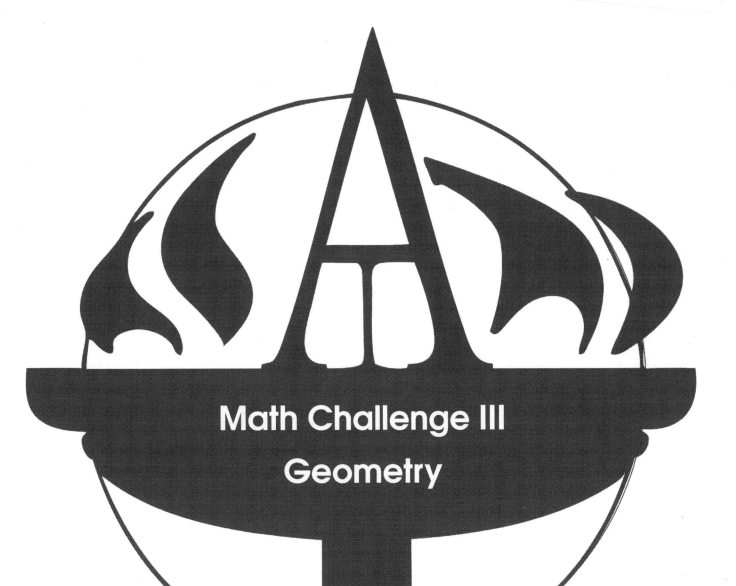

Math Challenge III
Geometry

Areteem Institute

Math Challenge III Geometry

Edited by   Kevin Wang
              John Lensmire
              David Reynoso
              Kelly Ren

Copyright © 2018 ARETEEM INSTITUTE

WWW.ARETEEM.ORG

PUBLISHED BY ARETEEM PRESS

ISBN: 1-944863-20-6
ISBN-13: 978-1-944863-20-3
First printing, August 2018.

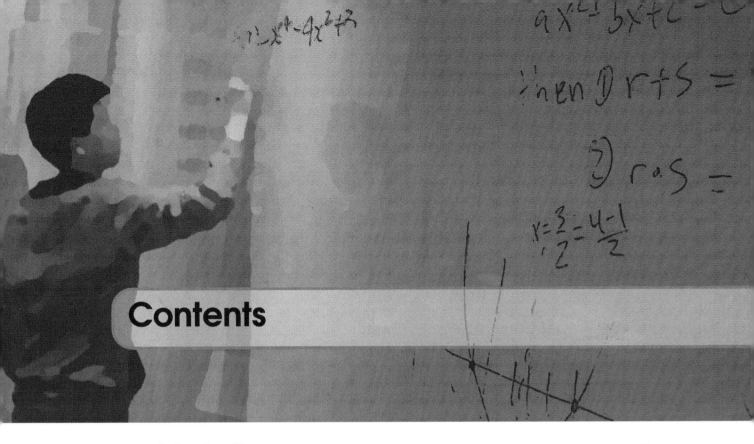

Contents

Solutions to the Example Questions 75

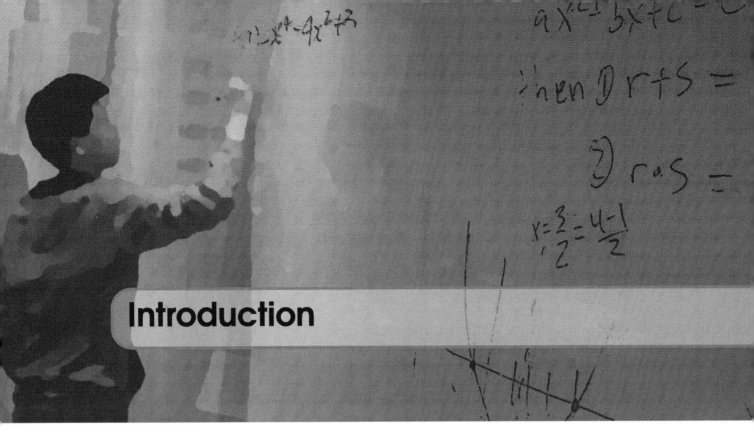

Introduction

The math challenge curriculum textbook series is designed to help students learn the fundamental mathematical concepts and practice their in-depth problem solving skills with selected exercise problems. Ideally, these textbooks are used together with Areteem Institute's corresponding courses, either taken as live classes or as self-paced classes. According to the experience levels of the students in mathematics, the following courses are offered:

- Fun Math Problem Solving for Elementary School (grades 3-5)
- Algebra Readiness (grade 5; preparing for middle school)
- Math Challenge I-A Series (grades 6-8; intro to problem solving)
- Math Challenge I-B Series (grades 6-8; intro to math contests e.g. AMC 8, ZIML Div M)
- Math Challenge I-C Series (grades 6-8; topics bridging middle and high schools)
- Math Challenge II-A Series (grades 9+ or younger students preparing for AMC 10)
- Math Challenge II-B Series (grades 9+ or younger students preparing for AMC 12)
- Math Challenge III Series (preparing for AIME, ZIML Varsity, or equivalent contests)
- Math Challenge IV Series (Math Olympiad level problem solving)

These courses are designed and developed by educational experts and industry professionals to bring real world applications into the STEM education. These programs are ideal for students who wish to win in Math Competitions (AMC, AIME, USAMO, IMO,

ARML, MathCounts, Math League, Math Olympiad, ZIML, etc.), Science Fairs (County Science Fairs, State Science Fairs, national programs like Intel Science and Engineering Fair, etc.) and Science Olympiad, or purely want to enrich their academic lives by taking more challenges and developing outstanding analytical, logical thinking and creative problem solving skills.

The Math Challenge III (MC III) courses are for students who are qualified to participate in the AIME contest, or at the equivalent level of experience. The MC III topics includes everything covered in MC-II with more depth, and the focus is more on various problem solving strategies, including pairing, change of variables, problem solving with logarithms, complex numbers, advanced techniques in number theory and combinatorics, advanced probability theory and techniques, geometric transformations, trigonometry, etc. The curricula have been proven to help students develop strong problem solving skills that make them perform well in math contests such as AIME, ZIML, and ARML.

The course is divided into four terms:

- Summer, covering Algebra
- Fall, covering Geometry
- Winter, covering Combinatorics
- Spring, covering Number Theory

The book contains course materials for Math Challenge III: Geometry.

We recommend that students take all four terms. Each of the individual terms is self-contained and does not depend on other terms, so they do not need to be taken in order, and students can take single terms if they want to focus on specific topics.

Students can sign up for the course at `classes.areteem.org` for the live online version or at `edurila.com` for the self-paced version.

About Areteem Institute

Areteem Institute is an educational institution that develops and provides in-depth and advanced math and science programs for K-12 (Elementary School, Middle School, and High School) students and teachers. Areteem programs are accredited supplementary programs by the Western Association of Schools and Colleges (WASC). Students may attend the Areteem Institute in one or more of the following options:

- Live and real-time face-to-face online classes with audio, video, interactive online whiteboard, and text chatting capabilities;
- Self-paced classes by watching the recordings of the live classes;
- Short video courses for trending math, science, technology, engineering, English, and social studies topics;
- Summer Intensive Camps held on prestigious university campuses and Winter Boot Camps;
- Practice with selected free daily problems and monthly ZIML competitions at ziml.areteem.org.

Areteem courses are designed and developed by educational experts and industry professionals to bring real world applications into STEM education. The programs are ideal for students who wish to build their mathematical strength in order to excel academically and eventually win in Math Competitions (AMC, AIME, USAMO, IMO, ARML, MathCounts, Math Olympiad, ZIML, and other math leagues and tournaments, etc.), Science Fairs (County Science Fairs, State Science Fairs, national programs like Intel Science and Engineering Fair, etc.) and Science Olympiads, or for students who purely want to enrich their academic lives by taking more challenging courses and developing outstanding analytical, logical, and creative problem solving skills.

Since 2004 Areteem Institute has been teaching with methodology that is highly promoted by the new Common Core State Standards: stressing the conceptual level understanding of the math concepts, problem solving techniques, and solving problems with real world applications. With the guidance from experienced and passionate professors, students are motivated to explore concepts deeper by identifying an interesting problem, researching it, analyzing it, and using a critical thinking approach to come up with multiple solutions.

Thousands of math students who have been trained at Areteem have achieved top honors and earned top awards in major national and international math competitions, including Gold Medalists in the International Math Olympiad (IMO), top winners and qualifiers at the USA Math Olympiad (USAMO/JMO) and AIME, top winners at the

Zoom International Math League (ZIML), and top winners at the MathCounts National Competition. Many Areteem Alumni have graduated from high school and gone on to enter their dream colleges such as MIT, Cal Tech, Harvard, Stanford, Yale, Princeton, U Penn, Harvey Mudd College, UC Berkeley, or UCLA. Those who have graduated from colleges are now playing important roles in their fields of endeavor.

Further information about Areteem Institute, as well as updates and errata of this book, can be found online at `http://www.areteem.org`.

Acknowledgments

This book contains many years of collaborative work by the staff of Areteem Institute. This book could not have existed without their efforts. Huge thanks go to the Areteem staff for their contributions!

The examples and problems in this book were either created by the Areteem staff or adapted from various sources, including other books and online resources. Especially, some good problems from previous math competitions and contests such as AMC, AIME, ARML, MATHCOUNTS, and ZIML are chosen as examples to illustrate concepts or problem-solving techniques. The original resources are credited whenever possible. However, it is not practical to list all such resources. We extend our gratitude to the original authors of all these resources.

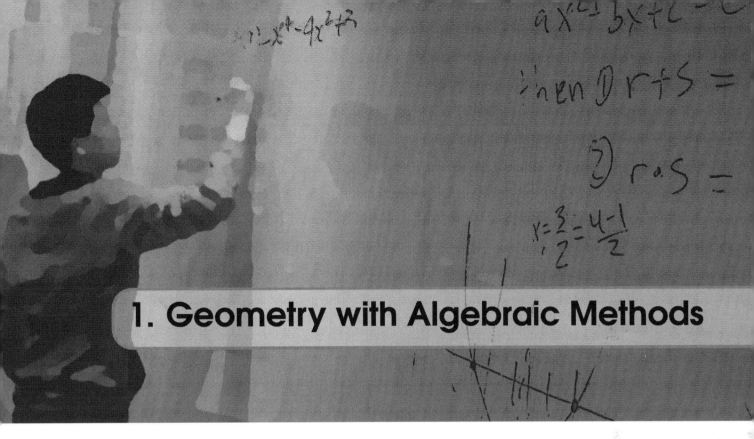

1. Geometry with Algebraic Methods

In geometry, the key for many proof problems is the measure of certain segments or angles. Some conclusions are given in the form of algebraic expressions. In addition, a lot of problems are calculation related. Using the geometric relations to set up equations and solving those equations is a very common method.

1.1 Example Questions

• Using linear equations

Problem 1.1 Given square $ABCD$, using $\overline{AB}$ as diameter and construct a semicircle inside the square. From C construct a tangent line $\overline{CF}$ to the semicircle with the tangent point E, intersecting $\overline{AD}$ at F. Find the ratio $DF : CD : CF$.

Problem 1.2 (AIME 1985) In triangle ABC, draw three lines from the vertices towards there respective opposite sides, passing through the same point, and split the triangle into six smaller triangles as shown. The areas of four of the smaller triangles are given in the diagram. Find the area of $\triangle ABC$.

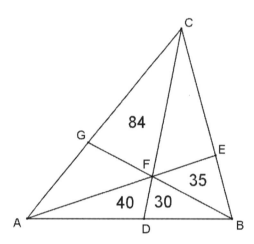

Problem 1.3 Given that $\odot O_2$ and $\odot O_3$ are externally tangent, and both of them are inside $\odot O_1$ and tangent to $\odot O_1$. Also $O_1O_2 = 3$, $O_1O_3 = 6$, and $O_2O_3 = 7$. Find the radii of these circles.

Problem 1.4 In right triangle ABC, $\angle C = 90°$, $AB = 13$, $AC = 12$. Circles $\odot O_1$ and $\odot O_2$ both have radius r, and $\odot O_1$ and $\odot O_2$ are externally tangent. Also given that $\odot O_1$ is tangent to $\overline{AB}$ and $\overline{AC}$ at M and G respectively, $\odot O_2$ is tangent to $\overline{AB}$ and $\overline{BC}$ at N and H respectively. Find the value of r.

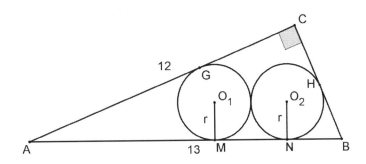

• Using quadratic equations

Problem 1.5 The lengths of the three sides of a triangle are three consecutive integers, and the largest angle is twice the smallest angle. Find the lengths of the sides.

Problem 1.6 As shown in the diagram, $\odot O_1$ and $\odot O_2$ intersect at A, B. Let $\overline{PQ}$ be their external common tangent line. Also, line $\overline{DT}$ is tangent to $\odot O_2$ at T, intersecting $\odot O_1$ at M, where M is also the midpoint of $\overline{DT}$. Let C and S be the intersection of $\overline{AB}$ with $\overline{DT}$ and $\overline{PQ}$ respectively. Find the ratios: (1) $SQ : SP$; (2) $CM : CT$.

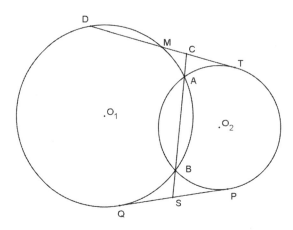

Problem 1.7 Let a, b, c be the side lengths of a triangle. How many real roots does the following equation have?

$$c^2 x^2 + (a^2 - b^2 - c^2)x + b^2 = 0$$

Problem 1.8 Square $ABCD$ has side length 6. Let M be the midpoint of $\overline{CD}$, and O be the circumcenter of $\triangle MAB$. Find the diameter of the incircle of $\triangle OAB$.

Problem 1.9 A trapezoid has area 32, and the sum of its two bases and altitude is 16. Also assume that one of the diagonals is perpendicular to the bases. Find the length of the other diagonal.

• Using other types of equations

Problem 1.10 One of the angles of a parallelogram is $60°$, and the ratio between the squares of the diagonals is $\dfrac{19}{7}$. Find the ratio of its two adjacent sides.

Problem 1.11 In a right triangle, the sum of the lengths of two legs is a, and the altitude on the hypotenuse is b. Find the length of the hypotenuse.

Problem 1.12 Given triangle ABC, $AB = 37$, $AC = 58$. Use A as the center and AB as the radius to draw a circle, intersecting $\overline{BC}$ at D where D is between B and C. If BD and DC are both integers, find the length of $\overline{BC}$.

• Using parameters

Sometimes it is convenient to set some parameters as bridge to what we need, and in many cases those parameters don't have to be solved.

Problem 1.13 As in the figure, $\odot O$ has three chords, PP', QQ', RR'. Let A be the intersection of PP' and RR', B be the intersection of PP' and QQ', and C be the intersection of QQ' and RR'. Assume $AP = BQ = CR$, $AR' = BP' = CQ'$, show that $\triangle ABC$ is equilateral.

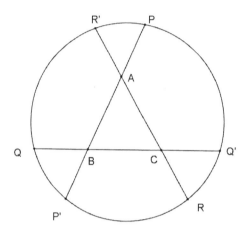

Problem 1.14 Let P be an interior point of square $ABCD$, $PA = 5$, $PD = 8$, and $PC = 13$. Find the area of square $ABCD$.

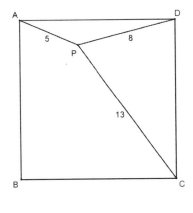

1.2 Practice Questions

Problem 1.15 Square $DEFG$ is inscribed in $\triangle ABC$, where $\overline{DE}$ is on $\overline{BC}$, F and G are on $\overline{AC}$ and $\overline{AB}$ respectively. Given that $BC = 60$, altitude $AH = 40$, find the side length of $DEFG$.

Problem 1.16 In $\triangle ABC$, $BC = 14$, $AC = 9$, and $AB = 13$, the incircle is tangent to sides $\overline{BC}$, $\overline{AC}$, and $\overline{AB}$ at D, E, and F respectively. Find the lengths of $\overline{AF}$, $\overline{BD}$, and $\overline{CE}$.

Problem 1.17 In trapezoid $ABCD$, $\overline{AD} \parallel \overline{BC}$, and let points E, G be on $\overline{AB}$, points F, H be on $\overline{DC}$ such that $AE = EG = GB$, and $DF = FH = HC$. Given that $AD = 20$, and $BC = 29$, find the lengths of $\overline{EF}$ and $\overline{GH}$.

Problem 1.18 The three sides of a triangle are three consecutive integers. The bisector of the largest angle splits the opposite side into two parts, where the shorter part has length $\dfrac{65}{9}$. Find the lengths of the sides.

Problem 1.19 In rectangle $ABCD$, the diagonal's length is 10. The inradii of $\triangle ABC$ and $\triangle ACD$ are both 2. Find the sides lengths of the rectangle and the distance between the centers of the two circles.

Problem 1.20 Let $\overline{AD}$, $\overline{AE}$, and $\overline{AF}$ be the altitude, angle bisector, and median of $\triangle ABC$, with lengths 24, $3\sqrt{65}$, and $4\sqrt{37}$ respectively. Find the length of $\overline{BC}$.

Problem 1.21 In trapezoid $ABCD$, $\overline{AD} \parallel \overline{BC}$, $\overline{AB} \perp \overline{BC}$. Let P be in the interior of $ABCD$, $PA = 1$, $PB = 2$, $PC = 3$, and $AB = BC = 2AD$. Find the area of $ABCD$.

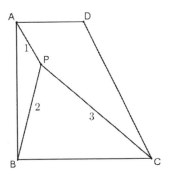

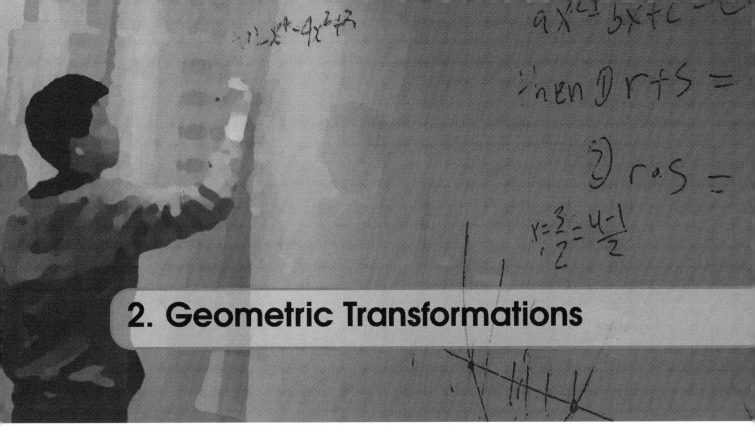

2. Geometric Transformations

A transformation in geometry, by definition, is a method that transforms one shape to another.

Some common transformations include **translation, rotation, and reflection**. These types of transformations only changes the positions of the objects, not the sizes or shapes, so they are called *congruence transformations*, or *isometries*. After a congruence transformation, the object retains its size, angles, area, and line lengths.

Transformations that change the size of the object but not the shape are called *dilations* or *compressions*, based on the factor of size change. A dilation based on a point is called a *homothety*.

Area-preserving transformation is also a very popular method. This transformation uses the following fact: if two triangles (or parallelograms) have equal bases and equal heights, they have the same area.

In this chapter we focus on the congruence transformations: translation, reflection, and rotation. Other common transformations are explored in the next chapter.

• Translation

A *translation* moves every point of a figure or a region by the same distance in a given direction.

Example 2.1

Villages A and B are located on the opposite sides of a river. People in both villages decide to build a bridge across the river to make it easier for them to visit each other. Assume the river banks are a pair of parallel lines, and the bridge XY should be perpendicular to the river banks, as shown in the diagram.

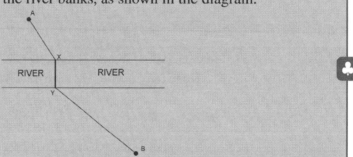

What is the best location X to build the bridge, so that the walk between the two villages is the shortest?

Solution

In order to find the shortest path, translate the location of Village B by towards the river by the distance that equals the width of the river, to point B'. Connect AB' with a straight line, intersecting the river bank (on the side of Village A) at point X'. Build the bridge across the river at X'. Let Y' be the other end of the bridge, then $A \to X' \to Y' \to B$ is the shortest path.

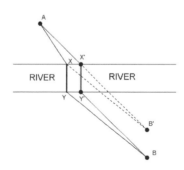

In order to show this path is indeed the shortest, note that $XY = X'Y'$ for any randomly picked X, and

$$XB' = YB, \quad X'B' = Y'B$$

because of the translation. Therefore (applying the Triangle Inequality)

$$
\begin{aligned}
AX + XY + YB &= AX + XB' + XY \\
&\geq AB' + X'Y' \\
&= AX' + X'B' + X'Y' \\
&= AX' + X'Y' + Y'B.
\end{aligned}
$$

Therefore X' is the best location to build the bridge.

• Reflection

Example 2.2

Villages A and B are located on the same side of a river, as shown in the diagram. Harry lives in Village A, and he plans to visit his friend Larry, who lives in Village B. Before visiting Larry, Harry also plans to fetch some water from the river.

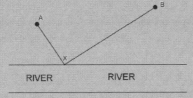

What is the best location X for Harry to get water, so that the entire walk path is the shortest?

Solution

Reflect the location of Village B across the near river bank to point B'. Connect AB' with a straight line, intersecting the river bank at point X'. Then $A \rightarrow X' \rightarrow B$ is the shortest path.

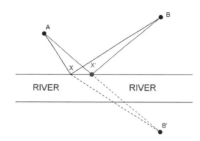

In order to show this path is the shortest, note for any randomly picked X,

$$XB' = XB, \quad X'B = X'B'$$

because of the reflection. Therefore (applying the Triangle Inequality)

$$
\begin{aligned}
AX + XB &= AX + XB' \\
&\geq AB' \\
&= AX' + X'B' \\
&= AX' + X'B.
\end{aligned}
$$

Therefore X' is the best location to fetch water.

• Rotation

Example 2.3

In regular hexagon $ABCDEF$, let M and N be the midpoints of $\overline{CD}$ and $\overline{DE}$ respectively, and $\overline{AM}$ and $\overline{BN}$ intersect at P. Compare the areas of triangle ABP and quadrilateral $MDNP$. Which area is bigger?

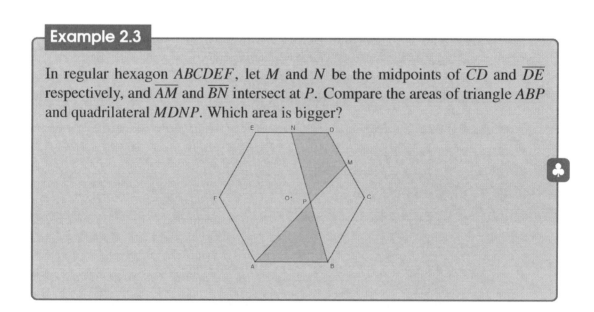

Solution

Rotate the hexagon $60°$ around its center O, so that points N, D, C, B become M, C, B, A respectively.

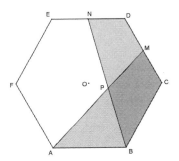

Therefore, the quadrilaterals $ABCM$ and $BCDN$ are congruent, and thus have the same area. Subtracting the common part $BCMP$ from both areas, we conclude that the areas of $\triangle ABP$ and quadrilateral $MDNP$ are equal.

2.1 Example Questions

Problem 2.1 Let $\overline{AB}$ and $\overline{CD}$ be two segments with length 1 and intersecting at E, and $\angle BED = 60°$. Show that $AC + BD \geq 1$.

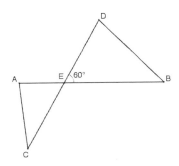

Problem 2.2 In hexagon $ABCDEF$, all 3 pairs of opposite sides are parallel. Also given that $AB - DE = CD - FA = EF - BC > 0$. Show that $ABCDEF$ is equiangular.

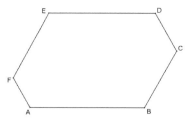

Problem 2.3 In rectangle $ABCD$, $AB = 3$, $BC = 4$. Fold the rectangle and flatten it so that A meets C. Find the area of the resulting overlap region.

Problem 2.4 In convex quadrilateral $ABCD$, let M be the midpoint of $\overline{BC}$, and $\angle AMD = 120°$. Show that $AB + \dfrac{1}{2}BC + CD \geq AD$.

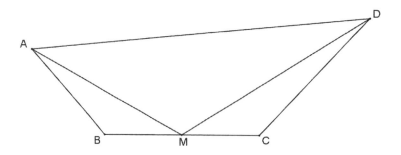

Problem 2.5 Let O be the center of a unit square. Let O be a vertex of a bigger square. What is the area of the overlap region of the two squares?

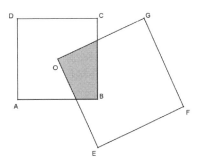

Problem 2.6 In quadrilateral $ABCD$, $\angle ADC = \angle ABC = 90°$, $AD = CD$, $\overline{DP} \perp \overline{AB}$ at P. Assume the area of $ABCD$ is 18, find DP.

Problem 2.7 The side length of square $ABCD$ is 1. Let P and Q be points on $\overline{AB}$ and $\overline{AD}$ respectively, and assume the perimeter of $\triangle APQ$ is 2. Find the measure of $\angle PCQ$ in degrees.

Problem 2.8 Given isosceles trapezoid $ABCD$, where $\overline{AD} \parallel \overline{BC}$. Construct adjacent parallelograms $ADEF$ and $CDEG$ outside the trapezoid $ABCD$. Show that $BD = FG$.

Problem 2.9 In $\triangle ABC$, let M be the midpoint of $\overline{BC}$. Two lines, both passing through M, and perpendicular to each other, intersect sides $\overline{AB}$ and $\overline{AC}$ at D and E respectively. Show that $BD + CE > DE$.

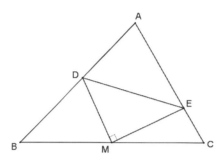

Problem 2.10 Let $\triangle ABC$ be equilateral triangle, and P be a point in the interior of $\triangle ABC$. Assume that $\angle APB = 115°$, $\angle BPC = 131°$. (1) Show that AP, BP, CP can be the three sides of a triangle; (2) find the degree measures of the angles of the triangle using AP, BP, CP as sides.

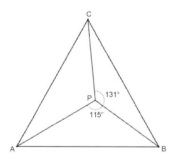

Problem 2.11 In isosceles triangle ABC, $\overline{BC}$ is the base, and D is the midpoint of $\overline{BC}$. Let E be a point in the interior of $\triangle ABD$. Show that $\angle AEB > \angle AEC$.

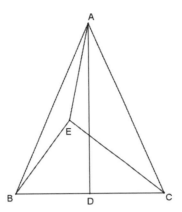

2.2 Practice Questions

Problem 2.12 Given $\triangle PAD$, and points B and C are on side $\overline{AD}$, where B is between A and C, and $AB = CD$. Show that $PA + PD > PB + PC$.

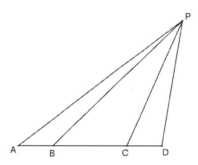

Problem 2.13 Let P be a point in the interior of rectangle $ABCD$, where $\angle PBC = \angle PDC$. Show that $PA \cdot PC + PB \cdot PD = AB \cdot AD$.

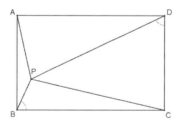

Problem 2.14 Let $ABCD$ be a square, and let P be a point on side $\overline{BC}$, and $\overline{AQ}$ bisect $\angle DAP$ and intersects $\overline{CD}$ at Q. Show that $AP = BP + DQ$.

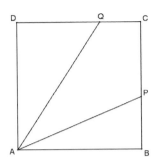

Problem 2.15 In $\triangle ABC$, find points P and Q on sides $\overline{AB}$ and $\overline{AC}$ respectively, so that $BQ + QP + PC$ is minimized.

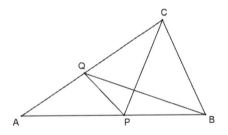

Problem 2.16 Let $\overline{AB}$ be a diameter of $\odot O$, and $\overline{DC}$ be a chord of $\odot O$ that is parallel to $\overline{AB}$. Let P be a point on $\overline{AB}$. Show that $PC^2 + PD^2 = PA^2 + PB^2$.

Problem 2.17 In $\triangle ABC$, $\overline{AD}$ is the altitude on side $\overline{BC}$. Given that $\angle A = 45°$, $BD = 2$, $DC = 3$, find the area of $\triangle ABC$.

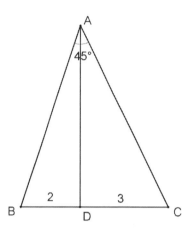

Problem 2.18 In acute triangle ABC, from the midpoint of each side, construct perpendicular lines to the other two sides. The six perpendicular lines form a hexagon. Let S be the area of this hexagon, and $[ABC]$ represent the area of $\triangle ABC$, compute the ratio $\dfrac{S}{[ABC]}$.

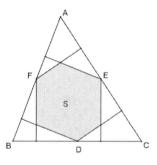

Problem 2.19 As shown in the diagram, $\overline{AB}$, $\overline{AC}$ are two chords of circle O, and $AB = AC$. Through C construct tangent line to $\odot O$, intersecting the extension of $\overline{BA}$ at D. Extend $\overline{CA}$, and $\overline{DE} \perp \overline{CA}$ at E. Find the ratio $CE : BD$.

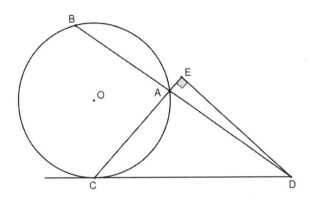

Problem 2.20 Let $\triangle ABC$ be an isosceles right triangle, $\angle C = 90°$. Let E and F be two points on $\overline{AB}$, where E is between A and F. Assume that $AE^2 + FB^2 = EF^2$, find the angle $\angle ECF$.

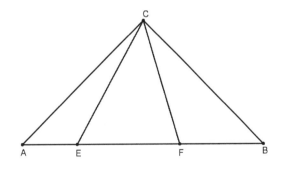

Problem 2.21 In $\triangle ABC$, D is the midpoint of $\overline{AB}$, E is a point on $\overline{AC}$, and line $\overline{DE}$ intersects the extension of $\overline{BC}$ at F. Show that $AE \cdot FC = BF \cdot EC$.

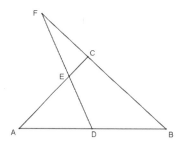

Problem 2.22 Let P be an interior point in square $ABCD$, and $PA = 1, PB = 3, PD = \sqrt{7}$. Find the area of $ABCD$.

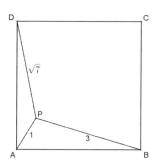

Problem 2.23 Let E be an arbitrary point in square $ABCD$ as shown, and the minimim possible value of $AE + BE + DE$ is $\sqrt{2} + \sqrt{6}$. Find the side length of the square.

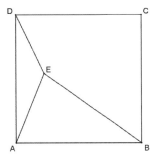

3. Transformations Continued

In this chapter we continue to explore the transformation method.

Transformations that change the size of the object but not the shape are called *dilations* or *compressions*, based on the factor of size change. A dilation based on a point is called a *homothety*.

Area-preserving transformation is also a very popular method. This transformation uses the following fact: if two triangles (or parallelograms) have equal bases and equal heights, they have the same area.

• Homothety

If the points $A', B', C', \ldots$ in one shape correspond one-to-one to the points $A, B, C, \ldots$ in another shape, satisfying

- The lines AA', BB', CC', $\ldots$ all pass through the same point O;
- $\dfrac{OA}{OA'} = \dfrac{OB}{OB'} = \dfrac{OC}{OC'} = \cdots = k (k \neq 0);$

then these two shapes are said to be *homothetic*. The point O is called the *homothetic center*, and k is the *homothetic ratio*. If $k = 1$, then the two shapes are congruent. The transformation between two homothetic shapes is called *homothetic transformation*, or simply *homothety*.

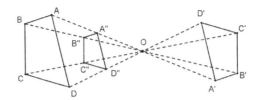

Properties of homothety:

- Two homothetic shapes must be similar, but not the other way around.
- Corresponding sides of homothetic shapes that do not pass through the homothetic center are parallel.
- In a homothetic transformation, the length ratio between corresponding segments is a fixed value.
- in a homothetic transformation, corresponding angles are congruent.

• Area-Preserving Transformations

The area-preserving transformations retain the area of the figure being transformed, without considering the shape and location of the figure. Main methods of area-preserving transformation:

- If two triangles (parallelograms) have equal base and equal height, the areas of the two triangles (parallelograms) are equal.
- Assume two triangles have a common base. If their remaining vertices fall on the same line parallel to the base, then the two triangles have the same area.
- Construct two triangles, $\triangle ABC$ and $\triangle A'BC$, where A and A' are on opposite sides of line $\overline{BC}$, so that the areas $[ABC] = [A'BC]$. Connect the vertices A and A', then $\overline{BC}$ bisects $\overline{AA'}$. The converse is also true: if $\overline{BC}$ bisects $\overline{AA'}$, then $[ABC] = [A'BC]$.

3.1 **Example Questions**

Problem 3.1 Let H and O be the orthocenter and circumcenter of triangle ABC. Let $\overline{ON} \perp \overline{BC}$ at N, find $\dfrac{AH}{ON}$.

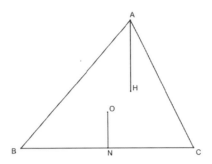

Problem 3.2 In triangle ABC, from vertex A construct lines perpendicular to the angle bisectors of $\angle ABC$, $\angle ACB$ and their exterior angles, with feet D, E, F, G respectively, as shown in the diagram. Are the four points D, E, F, G collinear?

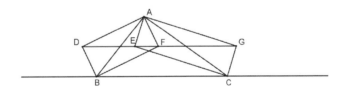

Problem 3.3 In acute triangle ABC, construct a rectangle $DEFG$, so that $\overline{DE}$ lies on $\overline{BC}$, and vertices G and F are on $\overline{AB}$ and $\overline{AC}$ respectively, satisfying $DE : EF = 1 : 2$.

Problem 3.4 Let A be a fixed point on a fixed circle $\odot O$. Point P is a point on a chord having A as one endpoint, and divides the chord into two segments with ratio $m : n$. Find the locus of P.

Problem 3.5 As shown in the diagram, three squares are lined up. The side lengths of the two smaller squares are 3 and 2 respectively. Find the area of $\triangle ABC$.

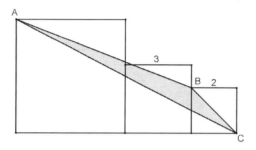

Problem 3.6 As shown in the diagram, in polygon $ABECD$, $\overline{AB} \parallel \overline{CD}$. Find a point C' on $\overline{CD}$ such that trapezoid $ABC'D$ has the same area as polygon $ABECD$.

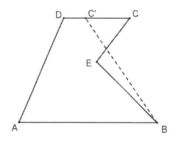

Problem 3.7 Construct a square that has the same area as a given trapezoid.

Problem 3.8 In trapezoid $ABCD$, $\overline{AD} \parallel \overline{BC}$, and M and N are midpoints of $\overline{AD}$ and $\overline{BC}$ respectively. Let P be an arbitrary point on $\overline{MN}$. Show that $[ABP] = [CDP]$.

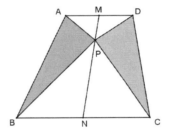

3.2   Practice Questions

The problems in this section include review problems of all types of transformations.

Problem 3.9 As shown, let D be the midpoint of a semicircle with diameter $\overline{AB}$. Given that $\overline{AC} \perp \overline{AB}$, $AC = AB$, and let E be the intersection of $\overline{CD}$ and $\overline{AB}$. Find the ratio of the areas $[BED] : [AEC]$.

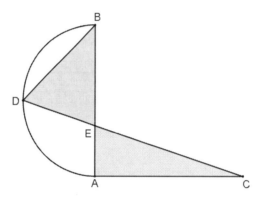

Problem 3.10 In $\triangle ABC$, $\angle ACB = 90°$, $AC = CB$. Line ℓ passes through C and $\ell \parallel \overline{AB}$. Let D be a point on ℓ so that $AD = AB$. Find all possible values of $\angle BAD$ in degrees.

Problem 3.11 In $\triangle ABC$, let E be the midpoint of sides $\overline{AC}$ and D be the point on $\overline{BC}$ such that $DC = 2BD$. Let F be the intersection of AD and BE. Given that $[BDF] = 1$, find the area of quadrilateral $DCEF$.

Problem 3.12 In convex quadrilateral $ABCD$, M and N are the midpoints of diagonals $\overline{AC}$ and $\overline{BD}$ respectively. Assume that the extensions of $\overline{AD}$ and $\overline{BC}$ intersect at P. Given that $[ABCD] = 100$, find the area $[PMN]$.

Problem 3.13 Three congruent circles $\odot O_1, \odot O_2, \odot O_3$ pass through a common point P, and all three circles are contained in a given triangle ABC, where $\odot O_1$ is tangent to $\overline{AB}$ and $\overline{AC}$, $\odot O_2$ is tangent to $\overline{AB}$ and $\overline{BC}$, $\odot O_3$ is tangent to $\overline{AC}$ and $\overline{BC}$,. Show that the incenter and circumcenter of the triangle and point P are collinear.

Problem 3.14 Seven straight lines pairwise intersect, and many angles are formed. Show that at least one of the angles is less than $26°$.

Problem 3.15 Given $\triangle ABC$ with area 10, from the three sides construct squares $ABDE$, $CAFG$, and $BCHK$ on the outside of $\triangle ABC$. Connect $\overline{EF}, \overline{GH}$, and $\overline{KD}$.

  1. Show that the three lengths EF, GH, KD can form a triangle;
  2. Find the area of that triangle.

Problem 3.16 In trapezoid $ABCD$, $\overline{AD} \parallel \overline{BC}$. Given that $BC = 10$, $CD = 5.5$. If $\angle ABC = 50°$, and $\angle ADC = 100°$, find the length of $\overline{AD}$.

Problem 3.17 In trapezoid $ABCD$, $\overline{AB} \parallel \overline{CD}$. Let E be the intersection of the two diagonals, M and N be the midpoints of sides $\overline{AD}$ and $\overline{BC}$ respectively, and also $MN = 6.5, AC = 12, BD = 5$. Find the measure of $\angle AED$.

Problem 3.18 In parallelogram $ABCD$, let E be the midpoint of $\overline{BC}$, and G be the intersection of $\overline{AE}$ and $\overline{BD}$. Suppose $[BEG] = 1$, find the area of parallelogram $ABCD$.

Problem 3.19 In hexagon $ABCDEF$, $AB = BC = CD = DE = EF = FA$, and $\angle A + \angle C + \angle E = \angle B + \angle D + \angle F$. Given that $[ABCDEF] = 6$, find $[BDF]$.

Problem 3.20 Given an angle $\angle POQ = 20°$. Let A be a point on $\overrightarrow{OQ}$ and B be a point on $\overrightarrow{OP}$, where $OA = 1, OB = 2$. Pick an arbitrary point A_1 on $\overline{OB}$ and another arbitrary point B_1 on $\overline{AQ}$. Let $k = AA_1 + A_1B_1 + B_1B$, find the minimum possible value of k.

Problem 3.21 In rectangle $ABCD$, $AB = 20$, $BC = 10$. Choose points M and N on $\overline{AC}$ and $\overline{AB}$ respectively, so that $BM + MN$ is minimized. Find this minimum value.

Problem 3.22 In square $ABCD$, let E be the midpoint of $\overline{CD}$, G be the intersection of $\overline{AE}$ and $\overline{BD}$. Let F be the point on $\overline{BC}$ such that $BF = \dfrac{1}{2}FC$. Find the angle AGF.

Problem 3.23 Let P be a point in the interior of equilateral triangle ABC. Suppose $\overline{PD} \perp \overline{BC}$ at D, $\overline{PE} \perp \overline{CA}$ at E, and $\overline{PF} \perp \overline{AB}$ at F.

  1. Show that the lengths PA, PB, PC can form a triangle $\mathscr{T}$;
  2. Show that the ratio between the area of the triangle $\mathscr{T}$ and $[DEF]$ is a fixed value;
  3. Find that fixed value.

Problem 3.24 As shown in the diagram, in isosceles right triangle ABC, $\angle ACB = 90°$. Points D and E are on side $\overline{AB}$, $AD = 1.7$, $BE = 2.64$, and $\angle DCE = 45°$. Find the length of $\overline{DE}$.

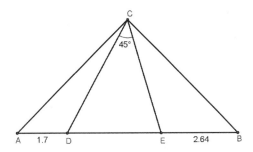

Problem 3.25 In square $ABCD$, points E, F, and G are the midpoints of $\overline{BC}$, $\overline{AE}$, and $\overline{DF}$ respectively. Given that the area of triangle BDG is 1, find the length of $\overline{AB}$.

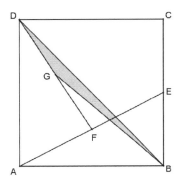

Problem 3.26 Let O be an interior point of convex quadrilateral $ABCD$, where $\angle AOB = \angle COD = 120°$. Also given that $AO = OB$ and $CO = OD$. Let K, L, and M be the midpoints of $\overline{AB}$, $\overline{BC}$, and $\overline{CD}$, respectively. Show that $\triangle KLM$ is equilateral.

Problem 3.27 Square $ABCD$ is divided into 4 small rectangles by two segments that are parallel to respective sides: $\overline{EF} \parallel \overline{AB}$ and $\overline{GH} \parallel \overline{BC}$. Also given that $[PFCG] = 2[AHPE]$, find the measure of $\angle GAF$.

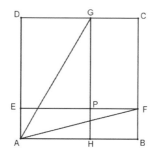

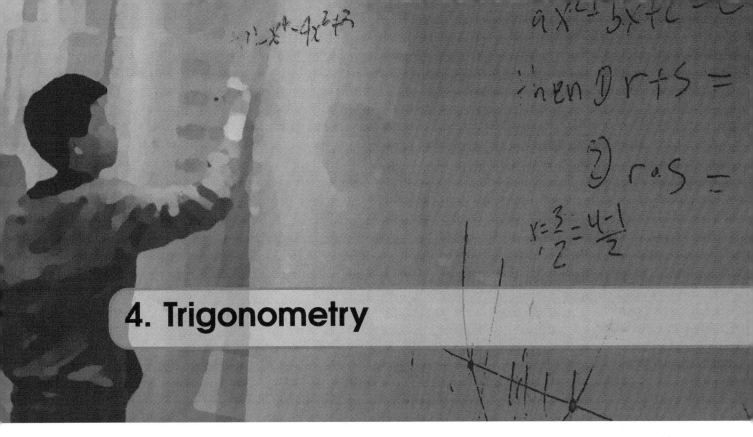

4. Trigonometry

Fundamental Trigonometry Formulas

I. The Trigonometry and Inverse Trigonometry Functions

Function	Domain	Range
$\sin x$	$(-\infty, +\infty)$	$[-1, 1]$
$\cos x$	$(-\infty, +\infty)$	$[-1, 1]$
$\tan x$	$x \neq \dfrac{\pi}{2} + k\pi$	$(-\infty, +\infty)$
$\cot x$	$x \neq k\pi$	$(-\infty, +\infty)$
$\sec x$	$x \neq \dfrac{\pi}{2} + k\pi$	$(-\infty, -1] \cup [1, +\infty)$
$\csc x$	$x \neq k\pi$	$(-\infty, -1] \cup [1, +\infty)$
$\arcsin x$	$[-1, 1]$	$\left[-\dfrac{\pi}{2}, \dfrac{\pi}{2}\right]$
$\arccos x$	$[-1, 1]$	$[0, \pi]$
$\arctan x$	$(-\infty, +\infty)$	$\left(-\dfrac{\pi}{2}, \dfrac{\pi}{2}\right)$
$\operatorname{arccot} x$	$(-\infty, +\infty)$	$(0, \pi)$

II. The Special Angles

Rad	0	$\pi/6$	$\pi/4$	$\pi/3$	$\pi/2$	$2\pi/3$	$3\pi/4$	$5\pi/6$	π
Deg	0	30	45	60	90	120	135	150	180
$\sin x$	0	1/2	$\sqrt{2}/2$	$\sqrt{3}/2$	1	$\sqrt{3}/2$	$\sqrt{2}/2$	1/2	0
$\cos x$	1	$\sqrt{3}/2$	$\sqrt{2}/2$	1/2	0	$-1/2$	$-\sqrt{2}/2$	$-\sqrt{3}/2$	-1
$\tan x$	0	$\sqrt{3}/3$	1	$\sqrt{3}$	-	$-\sqrt{3}$	-1	$-\sqrt{3}/3$	0
$\cot x$	-	$\sqrt{3}$	1	$\sqrt{3}/3$	0	$-\sqrt{3}/3$	-1	$-\sqrt{3}$	-
$\sec x$	1	$2\sqrt{3}/3$	$\sqrt{2}$	2	-	-2	$-\sqrt{2}$	$-2\sqrt{3}/3$	-1
$\csc x$	-	2	$\sqrt{2}$	$2\sqrt{3}/3$	1	$2\sqrt{3}/3$	$\sqrt{2}$	2	-

III. The Magic Hexagon (Wheel) of Trigonometry Identities

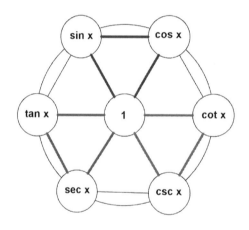

1. **On the diagonals—the reciprocal identities**

$$\sin x = \frac{1}{\csc x}, \qquad \cos x = \frac{1}{\sec x}, \qquad \tan x = \frac{1}{\cot x}$$

2. **On the upside-down triangles—the pythagorean identities**

$$\sin^2 x + \cos^2 x = 1$$
$$\tan^2 x + 1 = \sec^2 x$$
$$1 + \cot^2 x = \csc^2 x$$

3. **On the circle—the quotient identities** Any vertex is equal to the quotient of the next two consecutive vertices. It works in both directions.

$$\tan x = \frac{\sin x}{\cos x}, \qquad \sin x = \frac{\cos x}{\cot x}, \qquad \sec x = \frac{\tan x}{\sin x}, \qquad \text{etc.}$$

IV. Other Identities

1. **Co-functions (complementary angles)**

$$\sin\left(\frac{\pi}{2} - x\right) = \cos x, \qquad \cos\left(\frac{\pi}{2} - x\right) = \sin x, \qquad \tan\left(\frac{\pi}{2} - x\right) = \cot x$$

2. **Even-Odd properties**

$$\sin(-x) = -\sin x, \qquad \cos(-x) = \cos x, \qquad \tan(-x) = -\tan x$$

3. **Periodic properties**

$$\sin(x + 2\pi) = \sin x, \qquad \cos(x + 2\pi) = \cos x, \qquad \tan(x + \pi) = \tan x$$

4. **Supplementary angles**

$$\sin(\pi - x) = \sin x, \qquad \cos(\pi - x) = -\cos x, \qquad \tan(\pi - x) = -\tan x$$

5. **Other properties**

$$\sin(x + \pi) = -\sin x, \qquad \cos(x + \pi) = -\cos x$$

$$\sin(2\pi - x) = -\sin x, \qquad \cos(2\pi - x) = \cos x, \qquad \tan(2\pi - x) = -\tan x$$

6. **Sum & difference formulas**

$$\begin{aligned}
\sin(x \pm y) &= \sin x \cos y \pm \cos x \sin y \\
\cos(x \pm y) &= \cos x \cos y \mp \sin x \sin y \\
\tan(x \pm y) &= \frac{\tan x \pm \tan y}{1 \mp \tan x \tan y}
\end{aligned}$$

7. **Double angle formulas**

$$\begin{aligned}
\sin 2x &= 2 \sin x \cos x \\
\cos 2x &= \cos^2 x - \sin^2 x \\
&= 2 \cos^2 x - 1 \\
&= 1 - 2 \sin^2 x \\
\tan 2x &= \frac{2 \tan x}{1 - \tan^2 x}
\end{aligned}$$

8. **Power-reducing/Half angle formulas**

$$\sin^2 x = \frac{1 - \cos 2x}{2}, \qquad \cos^2 x = \frac{1 + \cos 2x}{2}$$

$$\tan^2 x = \frac{1 - \cos 2x}{1 + \cos 2x}, \qquad \tan x = \frac{\sin 2x}{1 + \cos 2x} = \frac{1 - \cos 2x}{\sin 2x}.$$

9. **Product-to-sum formulas**

$$\sin x \cos y = \frac{1}{2}\left(\sin(x+y) + \sin(x-y)\right)$$

$$\cos x \sin y = \frac{1}{2}\left(\sin(x+y) - \sin(x-y)\right)$$

$$\cos x \cos y = \frac{1}{2}\left(\cos(x+y) + \cos(x-y)\right)$$

$$\sin x \sin y = -\frac{1}{2}\left(\cos(x+y) - \cos(x-y)\right)$$

Proof. Start with the sum and difference formulas for $\sin x$:

$$\sin(x+y) = \sin x \cos y + \cos x \sin y,$$
$$\sin(x-y) = \sin x \cos y - \cos x \sin y.$$

Adding these identities,

$$\sin(x+y) + \sin(x-y) = 2\sin x \cos y,$$

Thus

$$\sin x \cos y = \frac{1}{2}\left(\sin(x+y) + \sin(x-y)\right).$$

If we apply subtraction instead of addition,

$$\sin(x+y) - \sin(x-y) = 2\cos x \sin y,$$

Hence

$$\cos x \sin y = \frac{1}{2}\left(\sin(x+y) - \sin(x-y)\right).$$

The remaining two identities are similarly proven using the sum and difference formulas for $\cos x$. ∎

10. **Sum-to-product formulas**

$$\sin x + \sin y = 2\sin\left(\frac{x+y}{2}\right)\cos\left(\frac{x-y}{2}\right)$$

$$\sin x - \sin y = 2\cos\left(\frac{x+y}{2}\right)\sin\left(\frac{x-y}{2}\right)$$

$$\cos x + \cos y = 2\cos\left(\frac{x+y}{2}\right)\cos\left(\frac{x-y}{2}\right)$$

$$\cos x - \cos y = -2\sin\left(\frac{x+y}{2}\right)\sin\left(\frac{x-y}{2}\right)$$

Proof. Start with the Product-to-Sum formula with new variables

$$\sin A \cos B = \frac{1}{2}\left(\sin(A+B) + \sin(A-B)\right).$$

Make a change of variables:

$$x = A+B, \quad y = A-B,$$

then

$$A = \frac{x+y}{2}, \quad B = \frac{x-y}{2}.$$

Substitute back into the formula,

$$\sin\left(\frac{x+y}{2}\right)\cos\left(\frac{x-y}{2}\right) = \frac{1}{2}(\sin x + \sin y),$$

which is

$$\sin x + \sin y = 2\sin\left(\frac{x+y}{2}\right)\cos\left(\frac{x-y}{2}\right).$$

The proofs of the other identities are all similar. ∎

11. **(Extended) Law of Sines**
   For a triangle ABC with circumradius R,

   $$\frac{a}{\sin A} = \frac{b}{\sin B} = \frac{c}{\sin C} = 2R$$

12. **Law of Cosines**
   For a triangle ABC,
   $$a^2 = b^2 + c^2 - 2bc\cos A$$
   $$b^2 = c^2 + a^2 - 2ca\cos B$$
   $$c^2 = a^2 + b^2 - 2ab\cos C$$

4.1 Example Questions

Problem 4.1 Assume the variable x in each question is in the valid domain.

(a) $\arcsin x + \arccos x = ?$

(b) $\arccos x + \arccos(-x) = ?$

Problem 4.2 Evaluate the following:

(a) $\sin 405°$

(b) $\cos 225°$

(c) $\sin 75°$

(d) $\cot 67.5°$

(e) $\cos\left(2\arcsin\dfrac{3}{5}\right)$

(f) $\sin\left(2\arccos\dfrac{4}{5}\right)$

Problem 4.3 Compute the following values.

(a) $\cos^4 \dfrac{\pi}{24} - \sin^4 \dfrac{\pi}{24}$

(b) $\cot 70° + 4\cos 70°$

(c) $\sin 10° \sin 30° \sin 50° \sin 70°$

(d) $\arctan \dfrac{1}{2} + \arctan \dfrac{1}{3}$

(e) $(1 - \cot 1°)(1 - \cot 44°)$

Problem 4.4 The quadratic equation $2x^2 - (\sqrt{3}+1)x + m = 0$ has two roots $\sin\theta$ and $\cos\theta$, find the value of $\dfrac{\sin\theta}{1 - \cot\theta} + \dfrac{\cos\theta}{1 - \tan\theta}$.

Problem 4.5 Given an equilateral triangle and its incircle. An arc on the incircle has the same length as the side length of the equilateral triangle. What is the measure, in radians, of the arc?

Problem 4.6 Given that α and β are acute angles, satisfying

$$\cos\alpha + \cos\beta - \cos(\alpha+\beta) = \frac{3}{2},$$

find the sum of all possible values of $\alpha + \beta$ in degrees. (ZIML Varsity March 2018)

Problem 4.7 Let $0 < \theta < \pi$, find the maximum value of $\sin\frac{\theta}{2}(1+\cos\theta)$. (Do not use calculus methods)

Problem 4.8 In $\triangle ABC$, let a, b, c be the lengths of the sides opposite angles A, B, C respectively. If $c - a$ equals the altitude h on side $\overline{AC}$, find the value of $\sin\frac{C-A}{2} + \cos\frac{C+A}{2}$.

Problem 4.9 Given $\alpha \in \left(\frac{\pi}{4}, \frac{\pi}{2}\right)$, list the following in increasing order:

$$(\cos\alpha)^{\cos\alpha}, (\sin\alpha)^{\cos\alpha}, (\cos\alpha)^{\sin\alpha}.$$

Problem 4.10 Let α and β be acute angles and $\alpha + \beta = 90°$, also $\sin\alpha$ and $\sin\beta$ are the roots of the equation $2x^2 - 2\sqrt{2}x + c = 0$. Suppose α equals k degrees, What is the value of $c + k$? (ZIML Varsity March 2018)

Problem 4.11 Evaluate

$$\sin^2 80^\circ + \sin^2 40^\circ - \cos 50^\circ \cos 10^\circ,$$

express your answer in decimal, rounded to the nearest hundredth if necessary. (ZIML Varsity June 2018)

4.2 Practice Questions

Problem 4.12 Evaluate $\dfrac{3-\sin 70^\circ}{2-\cos^2 10^\circ}$.

Problem 4.13 Evaluate $\dfrac{\cos 20^\circ}{\cos 35^\circ \sqrt{1-\sin 20^\circ}}$.

Problem 4.14 Evaluate $\cos\dfrac{\pi}{15}\cos\dfrac{2\pi}{15}\cos\dfrac{4\pi}{15}\cos\dfrac{8\pi}{15}$.

Problem 4.15 Evaluate $\tan 20^\circ + \tan 40^\circ + \sqrt{3}\tan 20^\circ \tan 40^\circ$.

Problem 4.16 If $\dfrac{1+\tan\alpha}{1-\tan\alpha}=2017$, then evaluate $\dfrac{1}{\cos 2\alpha}+\tan 2\alpha$.

Problem 4.17 In $\triangle ABC$, let a,b,c be the lengths of the sides opposite angles A,B,C respectively. Given that $9a^2+9b^2-19c^2=0$, find the value of $\dfrac{\cot C}{\cot A+\cot B}$.

Problem 4.18 In $\triangle ABC$, let a,b,c be the lengths of the sides opposite angles A,B,C respectively, and $b\neq 1$. Suppose $\dfrac{C}{A}$ and $\dfrac{\sin B}{\sin A}$ are both solutions of equation $\log_{\sqrt{b}}x=\log_b(4x-4)$. Determine the shape of $\triangle ABC$.

Problem 4.19 Given that $x, y \in \left\{ -\dfrac{\pi}{4}, \dfrac{\pi}{4} \right\}$, $a \in \mathbb{R}$, and

$$\begin{cases} x^3 + \sin x - 2a & = & 0, \\ 4y^3 + \sin y \cos y + a & = & 0. \end{cases}$$

Determine $\cos(x + 2y)$.

Problem 4.20 In $\triangle ABC$, the three angles A, B, C form an arithmetic sequence. If $c - a$ equals the altitude h on side $\overline{AC}$, find the value of $\sin \dfrac{C - A}{2}$.

Problem 4.21 Let x, y, z be in radians, and each of them equals the cosine value of the sum of the other two, namely, $x = \cos(y + z), y = \cos(z + x), z = \cos(x + y)$. Show that $x = y = z$.

Problem 4.22 Let $A_1 A_2 \cdots A_n$ be a regular n-gon, where $n \geq 4$. Given that $\dfrac{1}{|A_1 A_2|} = \dfrac{1}{|A_1 A_3|} + \dfrac{1}{|A_1 A_4|}$, find n.

5. Menelaus and Ceva

Theorem 5.1 Menelaus' Theorem

Given $\triangle ABC$, let X, Y, Z be points on the sides $\overline{BC}, \overline{CA}, \overline{AB}$ or their extensions, respectively. Then the points X, Y, Z are collinear if and only if

$$\frac{BX}{XC} \cdot \frac{CY}{YA} \cdot \frac{AZ}{ZB} = 1.$$

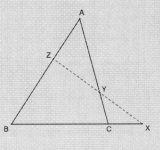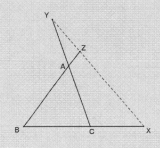

Remark

Sometimes the directions of the segments are taken into account when measuring the lengths, so the ratio like $\dfrac{BX}{XC}$ is negative if the point X is outside the segment $\overline{BC}$; then among the ratios in Theorem 5, either exactly one is negative or all three are negative, so with directed length,

$$\frac{BX}{XC} \cdot \frac{CY}{YA} \cdot \frac{AZ}{ZB} = -1.$$

If directions are also considered in angles and areas, the ratios in the other theorems, including the trigonometric forms of Menelaus' Theorem and Ceva's Theorem, can also be negative. For simplicity, in this lecture we do not consider directions for the measures of lengths, angles, and areas, so the ratios are always positive.

Menelaus' Theorem can also take the following trigonometric forms.

Theorem 5.2 Menelaus' Theorem - 1st Trig Form

Given $\triangle ABC$, let X, Y, Z be points on the sides $\overline{BC}, \overline{CA}, \overline{AB}$ or their extensions, respectively. Then the points X, Y, Z are collinear if and only if

$$\frac{\sin \angle BAX}{\sin \angle XAC} \cdot \frac{\sin \angle CBY}{\sin \angle YBA} \cdot \frac{\sin \angle ACZ}{\sin \angle ZCB} = 1.$$

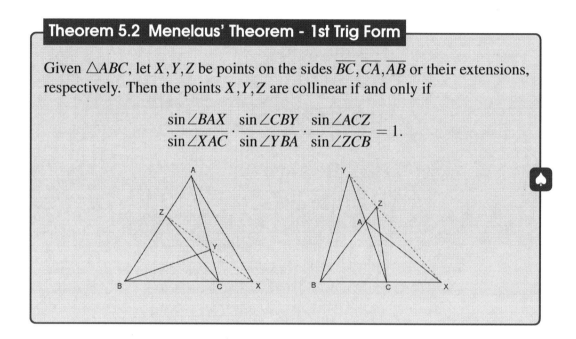

Theorem 5.3 Menelaus' Theorem - 2nd Trig Form

Given $\triangle ABC$, let X, Y, Z be points on the sides $\overline{BC}, \overline{CA}, \overline{AB}$ or their extensions, respectively. Let O be a point not on any of the lines $\overleftrightarrow{AB}, \overleftrightarrow{BC}, \overleftrightarrow{CA}$. Then the points X, Y, Z are collinear if and only if

$$\frac{\sin \angle BOX}{\sin \angle XOC} \cdot \frac{\sin \angle COY}{\sin \angle YOA} \cdot \frac{\sin \angle AOZ}{\sin \angle ZOB} = 1.$$

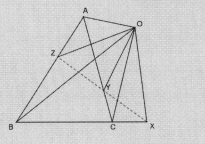

While Menelaus's Theorem is about the collinearity of three points, Ceva's Theorem is about the concurrency of three lines.

Theorem 5.4 Ceva's Theorem

Given $\triangle ABC$, let X, Y, Z be points on the sides $\overline{BC}, \overline{CA}, \overline{AB}$ or their extensions, respectively. Then the lines $\overline{AX}, \overline{BY}, \overline{CZ}$ are either concurrent or parallel if and only if

$$\frac{BX}{XC} \cdot \frac{CY}{YA} \cdot \frac{AZ}{ZB} = 1.$$

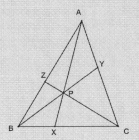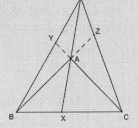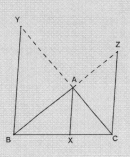

Ceva's Theorem also has trigonometric form.

Theorem 5.5 Ceva's Theorem - Trig Form

Given $\triangle ABC$, let X, Y, Z be points on the sides $\overline{BC}, \overline{CA}, \overline{AB}$ or their extensions, respectively. Then the lines $\overline{AX}, \overline{BY}, \overline{CZ}$ are either concurrent or parallel if and only if

$$\frac{\sin \angle BAX}{\sin \angle XAC} \cdot \frac{\sin \angle CBY}{\sin \angle YBA} \cdot \frac{\sin \angle ACZ}{\sin \angle ZCB} = 1.$$

Example 5.1

Given $\triangle ABC$, let A_1, B_1, C_1 be points on the arcs $\widehat{BC}, \widehat{CA}, \widehat{AB}$ of circumcircle of $\triangle ABC$, respectively. Then $\overline{AA_1}, \overline{BB_1}, \overline{CC_1}$ are concurrent if and only if

$$\frac{BA_1}{A_1C} \cdot \frac{CB_1}{B_1A} \cdot \frac{AC_1}{C_1B} = 1.$$

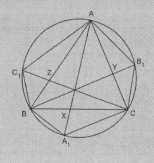

Solution

As shown in the diagram, let X be the intersection of $\overline{AA_1}$ and $\overline{BC}$, Y be the intersection of $\overline{BB_1}$ and $\overline{CA}$, Z be the intersection of $\overline{CC_1}$ and $\overline{AB}$. Let R be the circumradius of $\triangle ABC$, then by the Extended Law of Sines,

$$\frac{BA_1}{A_1C} = \frac{2R\sin \angle BAA_1}{2R\sin \angle A_1AC} = \frac{\sin \angle BAX}{\sin \angle XAC};$$

similarly,

$$\frac{CB_1}{B_1A} = \frac{\sin \angle CBY}{\sin \angle YBA}, \quad \frac{AC_1}{C_1B} = \frac{\sin \angle ACZ}{\sin \angle ZCB}.$$

Multiply the three fractions, and apply the Trigonometric Form of Ceva's Theorem, and the conclusion is proved.

5.1 Example Questions

Problem 5.1 In $\triangle ABC$, let D and E be points on $\overline{BC}$ and $\overline{AB}$ respectively, such that $\dfrac{BD}{DC} = \dfrac{1}{3}$ and $\dfrac{AE}{EB} = \dfrac{2}{3}$. Also let G be on $\overline{AD}$ such that $\dfrac{AG}{GD} = \dfrac{1}{2}$. Let F be the intersection of $\overline{EG}$ and $\overline{AC}$. Find $\dfrac{AF}{FC}$.

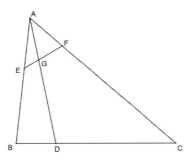

Problem 5.2 In parallelogram $ABCD$, let E and F be the midpoints of $\overline{AB}$ and $\overline{BC}$ respectively. $\overline{AF}$ and $\overline{CE}$ intersect at G, and $\overline{AF}$ and $\overline{DE}$ intersect at H. Find the ratio $AH : HG : GF$.

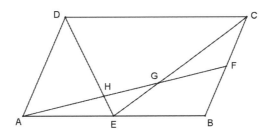

Problem 5.3 (AIME 1989) Let P be a point in the interior of $\triangle ABC$. Connect and extend $\overline{AP}, \overline{BP}, \overline{CP}$ and intersect $\overline{BC}, \overline{CA}, \overline{AB}$ at D, E, F respectively. Given that $AP = 6, BP = 9, PD = 6, PE = 3, CF = 20$, find the area of $\triangle ABC$.

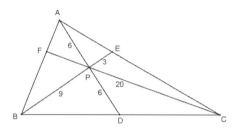

Problem 5.4 As shown in the diagram, in quadrilateral $ABCD$, the area ratio of $\triangle ABD, \triangle BCD$, and $\triangle ABC$ is $3 : 4 : 1$. Let M and N be points on $\overline{AC}$ and $\overline{CD}$ respectively, satisfying $\dfrac{AM}{AC} = \dfrac{CN}{CD} = r$, and B, M, N are collinear. Find the value of r.

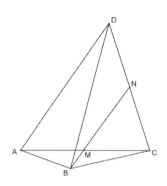

Problem 5.5 In quadrilateral $ABCD$, the diagonal $\overline{AC}$ bisects $\angle BAD$. Let E be a point on $\overline{CD}$, connect $\overline{BE}$ and intersect $\overline{AC}$ at F, and connect and extend $\overline{DF}$ to intersect $\overline{BC}$ at G. Show that $\angle GAC = \angle EAC$.

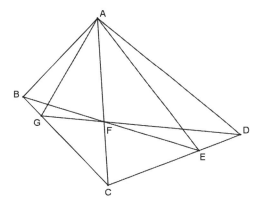

Problem 5.6 In $\triangle ABC$, D is a point on $\overline{BC}$ such that $\dfrac{BD}{DC} = \dfrac{1}{3}$. Let E be the midpoint of $\overline{AC}$, O be the intersection of $\overline{AD}$ and $\overline{BE}$, F be the intersection of $\overline{CO}$ and $\overline{AB}$. Find the ratio between the area of quadrilateral $BDOF$ and the area of $\triangle ABC$.

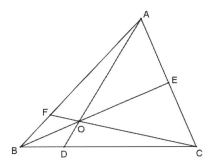

Problem 5.7 In $\triangle ABC$, $\angle ABC = \angle ACB = 40°$. Let P be an interior point in $\triangle ABC$, such that $\angle PAC = 20°$, $\angle PCB = 30°$. Find the degree measure of $\angle PBC$.

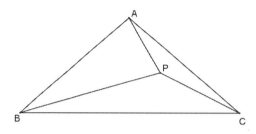

Problem 5.8 In $\triangle ABC$, $AB = AC$, $\angle A = 80°$. Let D be a point in the interior of $\triangle ABC$ such that $\angle DAB = \angle DBA = 10°$. Find the degree measure of $\angle ACD$.

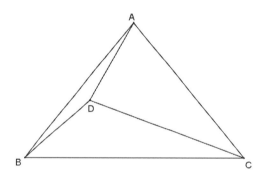

Problem 5.9 In $\triangle ABC$, $\angle BAC = 30°$, $\angle ABC = 70°$. Let M be a point in the interior of $\triangle ABC$ such that $\angle MAB = \angle MCA = 20°$. Find the degree measure of $\angle MBA$.

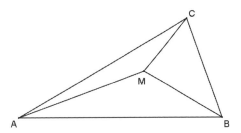

Problem 5.10 Let $\overline{AB}$ be a diameter of $\odot O$. Chord $\overline{CD}$ is perpendicular to $\overline{AB}$ at L, and points M and N are on segments $\overline{LB}$ and $\overline{LA}$ respectively, satisfying $LM : MB = LN : NA$. Rays $\overrightarrow{CM}$ and $\overrightarrow{CN}$ intersect $\odot O$ at E and F respectively. Show that $\overline{AE}$, $\overline{BF}$, and $\overline{OD}$ are concurrent.

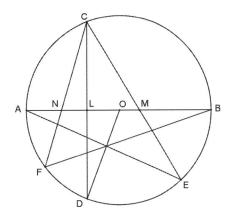

5.2 Practice Questions

Problem 5.11 Given $\triangle ABC$, let D and F be points on sides $\overline{AB}$ and $\overline{AC}$ respectively, and $AD : DB = CF : FA = 2 : 3$. Connect $\overline{DF}$ and intersect the extension of $\overline{BC}$ at E. Find $EF : FD$.

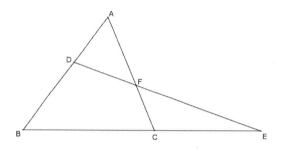

Problem 5.12 In $\triangle ABC$, points M and N trisect $\overline{AC}$ where M is between A and N; points X and Y trisect $\overline{BC}$ where X is between B and Y. $\overline{AY}$ intersects $\overline{BM}$ and $\overline{BN}$ at S and R respectively. Find the ratio between the areas of quadrilateral $SRNM$ and $\triangle ABC$.

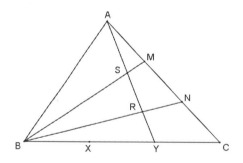

Problem 5.13 Let P be a point in the interior of parallelogram $ABCD$. Assume $\overline{MN}$ and $\overline{EF}$ pass through P, $\overline{MN} \parallel \overline{AD}$ and intersects $\overline{AB}$ and $\overline{CD}$ at M and N respectively, $\overline{EF} \parallel \overline{AB}$ and intersects $\overline{DA}$ and $\overline{BC}$ at E and F respectively. Show that $\overline{ME}$, $\overline{FN}$ and $\overline{BD}$ are concurrent.

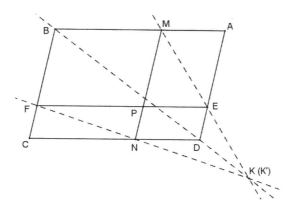

Problem 5.14 Given equilateral triangle ABC, let D, E, F be points on sides $\overline{BC}, \overline{CA}$ and $\overline{AB}$ respectively, such that $\dfrac{BD}{DC} = \dfrac{CE}{EA} = \dfrac{AF}{FB} = \dfrac{3}{n-3}$ where $n > 6$. Let $\triangle PQR$ be the triangle formed by the lines $\overline{AD}, \overline{BE},$ and $\overline{CF}$. Assume that $\dfrac{[PQR]}{[ABC]} = \dfrac{4}{49}$, find the value of n.

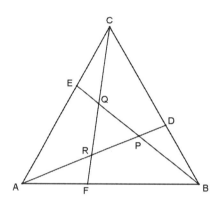

Problem 5.15 Let $ABCD$ be a square of side length 2. Let E be the midpoint of $\overline{AB}$, F be the midpoint of $\overline{BC}$. Given that $\overline{AF}$ intersects $\overline{DE}$ at I, and $\overline{BD}$ intersects $\overline{AF}$ at H. Find the area of quadrilateral $BEIH$.

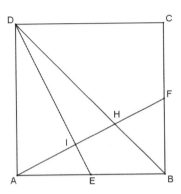

Problem 5.16 In $\triangle ABC$, $\angle ABC = 40°$, $\angle ACB = 20°$. Let N be a point inside $\triangle ABC$ such that $\angle NBC = 30°$ and $\angle NAB = 20°$. Find the degree measure of $\angle NCB$.

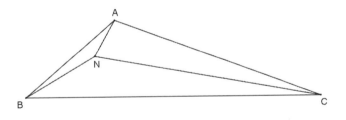

Problem 5.17 In $\triangle ABC$, $\angle ABC = \angle ACB = 40°$. Let P, Q be two points in the interior of $\triangle ABC$, such that $\angle PAB = \angle QAC = 20°$, and $\angle PCB = \angle QCA = 10°$. Show that B, P, Q are collinear.

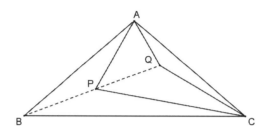

Problem 5.18 In $\triangle ABC$, $AB = AC$, $\angle A = 100°$. Let I be the incenter, and D be a point on $\overline{AB}$ such that $BD = BI$. Find the degree measure of $\angle BCD$.

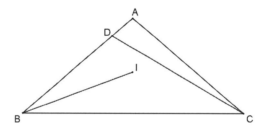

Problem 5.19 Given $\triangle ABC$, let X, Y and Z be points on the extension of $\overline{CB}, \overline{CA}$ and $\overline{BA}$ respectively. Also given that $\overline{XA}, \overline{YB}$ and $\overline{ZC}$ are tangent to the circumcircle of $\triangle ABC$. Show that X, Y, Z are collinear.

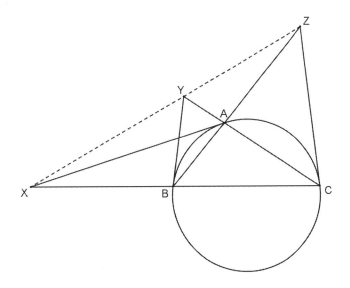

Problem 5.20 Let P be an arbitrary point in the interior of parallelogram $ABCD$. Through P construct line parallel to $\overline{AD}$ and intersect $\overline{AB}$ and $\overline{CD}$ at E and F respectively. Also, through P construct another line parallel to $\overline{AB}$, and intersect $\overline{AD}$ and $\overline{BC}$ at G and H respectively. Show that $\overline{AH}, \overline{CE}, \overline{DP}$ are concurrent.

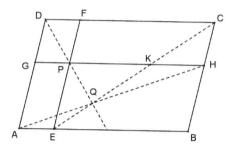

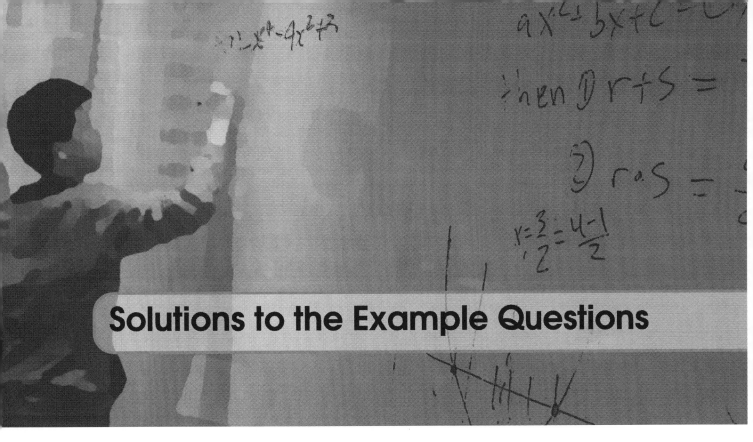

Solutions to the Example Questions

In the sections below you will find solutions to all of the Example Questions contained in this book.

Practice questions are meant to be used for homework, so their answers and solutions are not included. Teachers or math coaches may contact Areteem at info@areteem.org for answer keys and options for purchasing a Teachers' Edition of the course.

1   Solutions to Chapter 1 Examples

• Using linear equations

Problem 1.1 Given square $ABCD$, using $\overline{AB}$ as diameter and construct a semicircle inside the square. From C construct a tangent line $\overline{CF}$ to the semicircle with the tangent point E, intersecting $\overline{AD}$ at F. Find the ratio $DF : CD : CF$.

Answer

$3 : 4 : 5$

Solution

Let $AB = 1, AF = FE = x$. Then $DF = 1 - x, CF = CE + EF = 1 + x$, as shown.

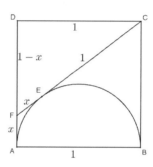

In the right triangle CDF,
$$DF^2 + CD^2 = CF^2,$$

so
$$(1-x)^2 + 1^2 = (1+x)^2.$$

Solve and get $x = \dfrac{1}{4}$, so $DF = \dfrac{3}{4}$, and $CF = \dfrac{5}{4}$. Thus
$$DF : CD : CF = 3 : 4 : 5.$$

Problem 1.2 (AIME 1985) In triangle ABC, draw three lines from the vertices towards there respective opposite sides, passing through the same point, and split the triangle into six smaller triangles as shown. The areas of four of the smaller triangles are given in the diagram. Find the area of $\triangle ABC$.

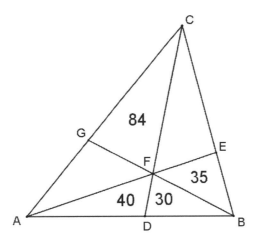

315

Label the points: let F be the intersecting point of three cevians (which means segment from one vertex to a point on the opposite side). Let D be the intersection of $\overline{CF}$ and $\overline{AB}$. Also label the unknown region to the lower left hand side x, and upper right side y. Based on areas,

$$\frac{AD}{DB} = \frac{40}{30} = \frac{84 + x + 40}{y + 35 + 30}$$

Similarly, $\dfrac{40 + 30 + 35}{x + 84 + y} = \dfrac{35}{x}$. Solve and get $x = 56, y = 70$. So

$$[ABC] = 84 + 56 + 40 + 30 + 85 + 70 = 315.$$

Problem 1.3 Given that $\odot O_2$ and $\odot O_3$ are externally tangent, and both of them are inside $\odot O_1$ and tangent to $\odot O_1$. Also $O_1 O_2 = 3$, $O_1 O_3 = 6$, and $O_2 O_3 = 7$. Find the radii of these circles.

8, 5, and 2

Let the radii of circles $\odot O_1, \odot O_2, \odot O_3$ be x, y, z respectively, as shown.

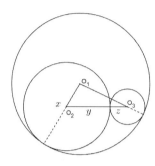

Then $x - y = 3, x - z = 6, y + z = 7$. Solve and get $x = 8, y = 5, z = 2$.

Problem 1.4 In right triangle ABC, $\angle C = 90°$, $AB = 13$, $AC = 12$. Circles $\odot O_1$ and $\odot O_2$ both have radius r, and $\odot O_1$ and $\odot O_2$ are externally tangent. Also given that $\odot O_1$ is tangent to $\overline{AB}$ and $\overline{AC}$ at M and G respectively, $\odot O_2$ is tangent to $\overline{AB}$ and $\overline{BC}$ at N and H respectively. Find the value of r.

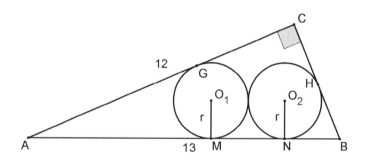

Answer

$26/17$

Solution 1

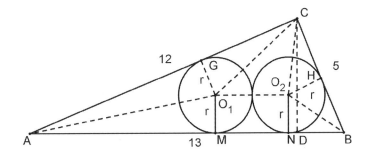

Connect $\overline{CO_1}, \overline{CO_2}, \overline{O_1O_2}, \overline{AO_1}, \overline{BO_2}$ as shown. We calculate the areas of the triangle in terms of r.

$$[ACO_1] = \frac{1}{2} \cdot 12 \cdot r = 6r,$$

$$[BCO_1] = \frac{1}{2} \cdot 5 \cdot r = \frac{5}{2}r,$$

$$[O_1O_2NM] = 2r^2,$$

$$[AMO_1] + [BNO_2] = \frac{1}{2}(13 - 2r)r = \frac{13}{2}r - r^2.$$

To calculate $[CO_1O_2]$, we first find the altitude $CD = \dfrac{2[ABC]}{13} = \dfrac{12 \times 5}{13} = \dfrac{60}{13}$, then the height of $\triangle CO_1O_2$ is $CD - r = \dfrac{60}{13} - r$. So

$$[CO_1O_2] = \frac{1}{2}2r \cdot \left(\frac{60}{13} - r\right) = \frac{60}{13}r - r^2.$$

Adding these areas together,

$$[ACO_1] + [BCO_1] + [O_1O_2NM] + [AMO_1] + [BNO_2] + [CO_1O_2] = [ABC],$$

$$6r + \frac{5}{2}r + 2r^2 + \frac{13}{2}r - r^2 + \frac{60}{13}r - r^2 = \frac{1}{2} \cdot 12 \cdot 5,$$

That is $\dfrac{255}{13}r = 30$, so $r = \dfrac{26}{17}$.

Solution 2

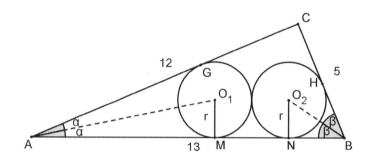

This solution uses trigonometry. As shown in the diagram, connect $\overline{AO_1}$ and $\overline{BO_2}$. Let $\angle CAB = 2\alpha$, and $\angle CBA = 2\beta$, then $\angle O_1AM = \alpha$, and $\angle O_2BN = \beta$. Calculate $BC = \sqrt{13^2 - 12^2} = 5$. Then

$$\sin 2\alpha = \frac{5}{13}, \quad \cos 2\alpha = \frac{12}{13}.$$

Similarly,

$$\sin 2\beta = \frac{12}{13}, \quad \cos 2\beta = \frac{5}{13}.$$

Using Half-Angle Formula to calculate $\tan \alpha$:

$$\tan \alpha = \frac{\sin 2\alpha}{1 + \cos 2\alpha} = \frac{\dfrac{5}{13}}{1 + \dfrac{12}{13}} = \frac{1}{5}.$$

Similarly we get

$$\tan \beta = \frac{\sin 2\beta}{1 + \cos 2\beta} = \frac{\dfrac{12}{13}}{1 + \dfrac{5}{13}} = \frac{2}{3}.$$

So

$$AM = \frac{r}{\tan \alpha} = 5r, \quad BN = \frac{r}{\tan \beta} = \frac{3}{2}r,$$

and

$$AM + MN + BN = 5r + 2r + \frac{3}{2}r = \frac{17}{2}r = AB = 13.$$

Solve the equation for r, then $r = \dfrac{26}{17}$.

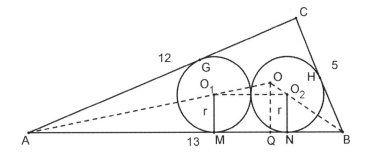

First calculate $BC = \sqrt{13^2 - 12^2} = 5$. Connect and extend $\overline{AO_1}$ and $\overline{BO_2}$, intersecting at O. Since $\overline{AO_2}$ and $\overline{AO_2}$ are angle bisectors, they must meet at the incenter of $\triangle ABC$, which is point O. From O construct the perpendicular line to side $\overline{BC}$, with foot Q. Then $\overline{OQ} \perp \overline{O_1O_2}$ as well. So $\overline{OQ}$ is the inradius of $\triangle ABC$, and since $\triangle ABC$ is a right triangle with hypotenuse $\overline{AB}$, the inradius equals

$$\frac{AC + BC - AB}{2} = \frac{12 + 5 - 13}{2} = 2,$$

which means $OQ = 2$.

Since $\overline{O_1O_2} \parallel \overline{AB}$, the two triangles $\triangle OO_1O_2 \sim \triangle OAB$. Thus

$$\frac{O_1O_2}{AB} = \frac{OQ - r}{OQ},$$

$$\frac{2r}{13} = \frac{2 - r}{2},$$

$$4r = 26 - 13r,$$

$$r = \frac{26}{17}.$$

• Using quadratic equations

Problem 1.5 The lengths of the three sides of a triangle are three consecutive integers, and the largest angle is twice the smallest angle. Find the lengths of the sides.

Assume $\angle A = 2\angle B$, and assume $AB = x$ is the middle number. So $AC = x - 1$ and $BC = x + 1$, as shown in the diagram.

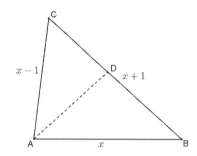

Let $\overline{AD}$ be the angle bisector at A. Then $\angle B = \angle BAD = \angle DAC$. Thus $\triangle CAD \sim \triangle CBA$, so $\dfrac{CD}{AC} = \dfrac{AC}{BC}$, hence $CD = \dfrac{AC^2}{BC}$. Also, since $\angle BAD = \angle DAC$, $\dfrac{CD}{BD} = \dfrac{AC}{AB}$, and then

$$\frac{CD}{BD+DC} = \frac{AC}{AB+AC},$$

so

$$\frac{AC^2}{BC^2} = \frac{AC}{AB+AC}.$$

Therefore

$$\frac{(x-1)^2}{(x+1)^2} = \frac{x-1}{2x-1}.$$

Now we solve for x. We know that $x \neq 1$, then

$$\frac{x-1}{(x+1)^2} = \frac{1}{2x-1},$$

which is

$$(x-1)(2x-1) = (x+1)^2,$$

and thus

$$2x^2 - 3x + 1 = x^2 + 2x + 1,$$
$$x^2 - 5x = 0.$$

Since $x \neq 0$, we get $x = 5$. Thus the three sides are $4, 5, 6$.

Problem 1.6 As shown in the diagram, $\odot O_1$ and $\odot O_2$ intersect at A, B. Let $\overline{PQ}$ be their external common tangent line. Also, line $\overline{DT}$ is tangent to $\odot O_2$ at T, intersecting $\odot O_1$ at M, where M is also the midpoint of $\overline{DT}$. Let C and S be the intersection of $\overline{AB}$ with $\overline{DT}$ and $\overline{PQ}$ respectively. Find the ratios: (1) $SQ : SP$; (2) $CM : CT$.

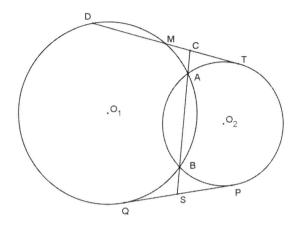

Answer

(1) $1:1$; (2) $1:2$.

Solution

By Power of a Point, $SQ^2 = SB \cdot SC = SP^2$, so $SQ = SP$, thus $SQ:SP = 1:1$.
Also, $CT^2 = CA \cdot CB$, and $CM \cdot CD = CA \cdot CB$, thus

$$TC^2 = CM \cdot CD = CM(CT + 2CM).$$

Let $CM = 1, CT = x$, then the equation becomes

$$x^2 = x + 2.$$

Solve and get $x = 2$ or $x = -1$. Since $x > 0$, we get $x = 2$.
Therefore $CM:CT = 1:2$.

Problem 1.7 Let a, b, c be the side lengths of a triangle. How many real roots does the following equation have?

$$c^2 x^2 + (a^2 - b^2 - c^2)x + b^2 = 0$$

Answer

No real roots.

Solution

Since a, b, c are the side lengths of a triangle, we have $a+b > c, b+c > a, c+a > b$. The discriminant of the equation is

$$
\begin{aligned}
(a^2 - b^2 - c^2)^2 - 4b^2c^2 &= (a^2 - b^2 - c^2 + 2bc)(a^2 - b^2 - c^2 - 2bc) \\
&= (a^2 - (b-c)^2)(a^2 - (b+c)^2) \\
&= (a+b-c)(a-b+c)(a-b-c)(a+b+c) \\
&< 0.
\end{aligned}
$$

Problem 1.8 Square $ABCD$ has side length 6. Let M be the midpoint of $\overline{CD}$, and O be the circumcenter of $\triangle MAB$. Find the diameter of the incircle of $\triangle OAB$.

Answer

2

Solution

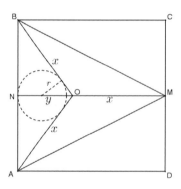

As shown in the diagram, let $OA = OB = OC = x$, and $ON = y$, then $x + y = 6$ and $x^2 - y^2 = 3^2$. Solve this system to get $x = \dfrac{15}{4}, y = \dfrac{9}{4}$. Let r be the inradius of $\triangle OAB$, and $2p$ be the perimeter of $\triangle OAB$, then

$$[OAB] = \frac{1}{2} \cdot 2p \cdot r = pr,$$

so

$$r = \frac{[OAB]}{p} = \frac{AB \cdot y/2}{(AB + 2x)/2} = \frac{6 \cdot \dfrac{9}{4}}{6 + 2 \cdot \dfrac{15}{4}} = 1.$$

Therefore the diameter of the incircle of $\triangle OAB$ is 2.

Problem 1.9 A trapezoid has area 32, and the sum of its two bases and altitude is 16. Also assume that one of the diagonals is perpendicular to the bases. Find the length of the other diagonal.

Answer

$8\sqrt{2}$

Solution

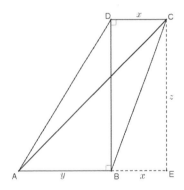

As shown in the diagram, let x and y be the bases of the trapezoid, and z be the altitude, then

$$\begin{cases} x+y+z & = & 16 \\ \dfrac{1}{2}z(x+y) & = & 32 \end{cases}$$

Solve and get $x+y=8$ and $z=8$. Through C construct $\overline{CE} \perp \overline{AB}$, where the foot E is on the extension of $\overline{AB}$. Then $AE = x+y = z = 8$, which means $\triangle ACE$ is an isosceles right triangle. Therefore $AC = \sqrt{AE^2 + CE^2} = 8\sqrt{2}$.

● **Using other types of equations**

Problem 1.10 One of the angles of a parallelogram is $60°$, and the ratio between the squares of the diagonals is $\dfrac{19}{7}$. Find the ratio of its two adjacent sides.

Answer

$\dfrac{3}{2}$ or $\dfrac{2}{3}$

Solution

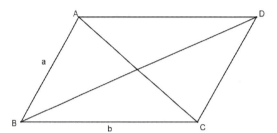

Assume the parallelogram is $ABCD$, where $\angle B = 60°$. Let $AB = a, BC = b$. Use the Law of Cosines on the diagonals, and set up equations:

$$AC^2 = a^2 + b^2 - 2ab\cos 60° = a^2 + b^2 - ab;$$

$$BD^2 = a^2 + b^2 - 2ab\cos 120° = a^2 + b^2 + ab.$$

Thus

$$\frac{a^2 + b^2 + ab}{a^2 + b^2 - ab} = \frac{19}{7}.$$

Since $b \neq 0$, divide the top and the bottom of the left hand side by b^2,

$$\frac{\left(\dfrac{a}{b}\right)^2 + 1 + \dfrac{a}{b}}{\left(\dfrac{a}{b}\right)^2 + 1 - \dfrac{a}{b}} = \frac{19}{7},$$

Let $x = \dfrac{a}{b}$, then

$$\begin{aligned}
\frac{x^2 + x + 1}{x^2 - x + 1} &= \frac{19}{7}, \\
7x^2 + 7x + 7 &= 19x^2 - 19x + 19, \\
12x^2 - 26x + 12 &= 0, \\
2(2x - 3)(3x - 2) &= 0.
\end{aligned}$$

Therefore $x = \dfrac{3}{2}$ or $\dfrac{2}{3}$, which means $\dfrac{a}{b} = \dfrac{3}{2}$ or $\dfrac{2}{3}$.

Problem 1.11 In a right triangle, the sum of the lengths of two legs is a, and the altitude on the hypotenuse is b. Find the length of the hypotenuse.

Answer

$\sqrt{a^2 + b^2} - b$

Solution

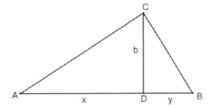

As shown in the diagram, we know $AC + BC = a$, $CD = b$. Let $AD = x$, $BD = y$, then by Pythagorean Theorem,
$$AC^2 + BC^2 = (x+y)^2.$$
Also the area $[ABC] = \dfrac{1}{2}(x+y)b = \dfrac{1}{2}AC \cdot BC$, we get
$$AC \cdot DC = (x+y)b.$$

Thus
$$a^2 = (AC+BC)^2 = AC^2 + BC^2 + 2AC \cdot BC = (x+y)^2 + 2(x+y)b,$$

therefore
$$(x+y)^2 + 2b(x+y) - a^2 = 0,$$

and
$$x+y = \frac{-2b \pm \sqrt{4b^2 + 4a^2}}{2} = -b \pm \sqrt{a^2 + b^2}.$$

Taking the positive value, the hypotenuse $x+y = \sqrt{a^2+b^2} - b$.

Problem 1.12 Given triangle ABC, $AB = 37$, $AC = 58$. Use A as the center and AB as the radius to draw a circle, intersecting $\overline{BC}$ at D where D is between B and C. If BD and DC are both integers, find the length of $\overline{BC}$.

Answer

57

Solution

Let E be the intersection of $\overline{AC}$ and the circle, and extend $\overline{CA}$ to intersect the circle at another point F, as shown.

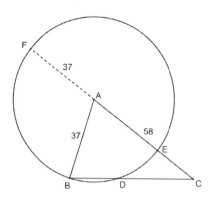

By Power of a Point,

$$CD \cdot CB = CE \cdot CF = (58+37)(58-37) = 95 \times 21 = 1995.$$

Knowing that CD and CB are both integers, $CB < CF = 95$, the only possibility is $CD = 35$ and $CB = 57$. So the answer is 57.

• Using parameters

Sometimes it is convenient to set some parameters as bridge to what we need, and in many cases those parameters don't have to be solved.

Problem 1.13 As in the figure, $\odot O$ has three chords, PP', QQ', RR'. Let A be the intersection of PP' and RR', B be the intersection of PP' and QQ', and C be the intersection of QQ' and RR'. Assume $AP = BQ = CR$, $AR' = BP' = CQ'$, show that $\triangle ABC$ is equilateral.

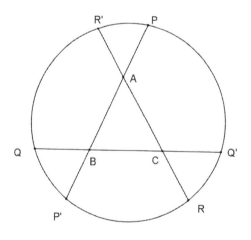

Solution

Label the segments as shown,

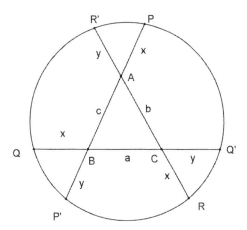

Set up equations using Power of a Point Theorem,

$$\begin{aligned}
x(y+c) &= y(x+b), \\
x(y+a) &= y(x+c), \\
x(y+b) &= y(x+a),
\end{aligned}$$

simplify to get

$$\begin{aligned}
xc &= yb, \\
xa &= yc, \\
xb &= ya,
\end{aligned}$$

multiplying all together,

$$x^3 \cdot abc = y^3 \cdot abc,$$

thus $x = y$, and then $a = b = c$. Therefore, $\triangle ABC$ is equilateral.

Problem 1.14 Let P be an interior point of square $ABCD$, $PA = 5$, $PD = 8$, and $PC = 13$. Find the area of square $ABCD$.

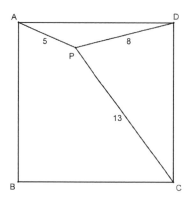

Solution 1

From P construct lines perpendicular to sides $\overline{AD}$ and $\overline{CD}$ with feet E and F respectively, as shown. Let $AB = a, PE = m, PF = n$.

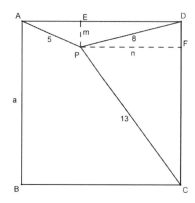

By Pythagorean Theorem,

$$\begin{cases} \sqrt{25 - m^2} + n &= a, \\ \sqrt{169 - n^2} + m &= a, \\ m^2 + n^2 &= 64. \end{cases}$$

Rearranging and squaring the first and second equations,

$$\begin{aligned} 25 - m^2 &= (a - n)^2, \\ 169 - n^2 &= (a - m)^2, \end{aligned}$$

expanding and rearranging,

$$\begin{aligned} 25 &= a^2 - 2an + n^2 + m^2, \\ 169 &= a^2 - 2am + m^2 + n^2, \end{aligned}$$

using the fact that $m^2 + n^2 = 64$, we get

$$
\begin{aligned}
a^2 + 39 &= 2an, \\
a^2 - 105 &= 2am,
\end{aligned}
$$

hence

$$
\begin{aligned}
a^2 + 39 &= 2an, \\
a^2 - 105 &= 2am,
\end{aligned}
$$

squaring,

$$
\begin{aligned}
(a^2 + 39)^2 &= 4a^2n^2, \\
(a^2 - 105)^2 &= 4a^2m^2,
\end{aligned}
$$

adding these equations and using the fact that $m^2 + n^2 = 64$ again,

$$
\begin{aligned}
(a^2 + 39)^2 + (a^2 - 105)^2 &= 256a^2, \\
a^4 + 78a^2 + 1521 + a^4 - 210a^2 + 11025 &= 256a^2, \\
2a^4 - 388a^2 + 12546 &= 0, \\
a^4 - 194a^2 + 6273 &= 0, \\
(a^2 - 41)(a^2 - 153) &= 0,
\end{aligned}
$$

thus $a^2 = 41$ or $a^2 = 153$. Since P is in the interior of $ABCD$, $CP < AC = \sqrt{2}a$, so $13 < \sqrt{2}a$, thus $a^2 > 84.5$, so the only possible value is $a^2 = 153$.
Therefore the area of $ABCD$ is 153.

Solution 2

We rotate $\triangle DAP$ about point D by $90°$ so that the side $\overline{AD}$ coincides with side $\overline{CD}$, and point P reaches point P', as shown. We know that $\triangle DP'P$ is an isosceles right triangle, and $\angle DP'P = 45°$. Let $\angle PP'C = \theta$.

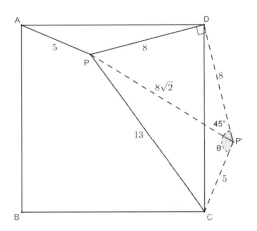

Applying Law of Cosines on $\triangle PCP'$,

$$13^2 = 5^2 + (8\sqrt{2})^2 - 2 \times 5 \times 8\sqrt{2}\cos\theta,$$

thus

$$\cos\theta = \frac{25 + 128 - 169}{2 \times 5 \times 8\sqrt{2}} = -\frac{\sqrt{2}}{10}.$$

Hence

$$\sin\theta = \sqrt{1 - \cos^2\theta} = \frac{7\sqrt{2}}{10}.$$

To calculate the area of $ABCD$, we use Law of Cosines again,

$$
\begin{aligned}
[ABCD] = CD^2 &= 8^2 + 5^2 - 2 \times 8 \times 5 \cos\angle DP'C \\
&= 89 - 80\cos(45° + \theta) \\
&= 89 - 80(\cos 45° \cos\theta - \sin 45° \sin\theta) \\
&= 89 - 80\left(\frac{\sqrt{2}}{2} \cdot \left(-\frac{\sqrt{2}}{10}\right) - \frac{\sqrt{2}}{2} \cdot \frac{7\sqrt{2}}{10}\right) \\
&= 89 + 8 + 56 \\
&= 153.
\end{aligned}
$$

2   Solutions to Chapter 2 Examples

Problem 2.1 Let $\overline{AB}$ and $\overline{CD}$ be two segments with length 1 and intersecting at E, and $\angle BED = 60°$. Show that $AC + BD \geq 1$.

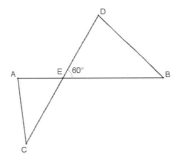

Translate $\overline{AC}$ to $\overline{BB'}$, connect $\overline{DB'}$, as shown.

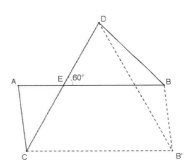

Clearly $CB' = AB = 1 = CD$, so $\triangle CDB'$ is equilateral, thus $DB' = 1$. By Triangle Inequality, $DB + BB' \geq DB' = 1$. And since $BB' = AC$, we get

$$AC + BD \geq 1.$$

Problem 2.2 In hexagon $ABCDEF$, all 3 pairs of opposite sides are parallel. Also given that $AB - DE = CD - FA = EF - BC > 0$. Show that $ABCDEF$ is equiangular.

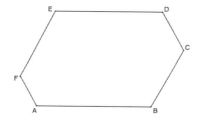

Solution

Construct the parallel line of $\overline{FE}$ through A, the parallel line of $\overline{BA}$ through C, and the parallel line of $\overline{DC}$ through E, as shown.

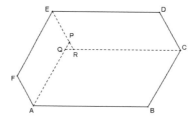

These new lines intersect at P, Q, R respectively, as shown in the diagram. Thus

$$PQ = AP - AQ = EF - BC.$$

Similarly $QR = AB - DE$ and $RP = CD - FA$. Since $AB - DE = CD - FA = EF - BC$, $\triangle PQR$ is equilateral. Therefore,

$$\angle PQR = \angle QRP = \angle RPQ = 60°.$$

Hence, $\angle PEF = 60°$ and $\angle F = 120°$. Similarly $\angle B = \angle D = 120°$, and $\angle DEP = 60°$. Then $\angle DEF = 60° + 60° = 120°$. Similarly $\angle BCD = \angle FAB = 120°$.

Therefore, the hexagon $ABCDEF$ is equiangular.

Problem 2.3 In rectangle $ABCD$, $AB = 3$, $BC = 4$. Fold the rectangle and flatten it so that A meets C. Find the area of the resulting overlap region.

Answer

75/16

Solution

As shown in the diagram, let $DE = x$, then $AE = EC = CF = 4 - x$.

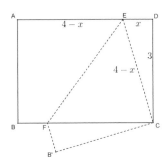

In the right triangle BED,

$$3^2 + x^2 = (4 - x)^2.$$

Solve for x, we get

$$9 + x^2 = 16 - 8x + x^2,$$

Therefore $x = 7/8$. So $CF = 4 - x = 25/8$, and the area of the overlap region is

$$[CEF] = \frac{1}{2}CF \cdot CD = \frac{1}{2} \cdot \frac{25}{8} \cdot 3 = \frac{75}{16}.$$

Problem 2.4 In convex quadrilateral $ABCD$, let M be the midpoint of $\overline{BC}$, and $\angle AMD = 120°$. Show that $AB + \frac{1}{2}BC + CD \geq AD$.

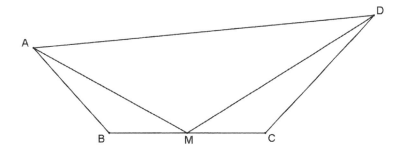

Solution

Reflect point B about $\overline{AM}$ to B', and reflect point C about $\overline{DM}$ to C', then connect $\overline{AB'}$, $\overline{B'M}$, $\overline{B'C'}$, $\overline{C'M}$, and $\overline{DC'}$, as shown.

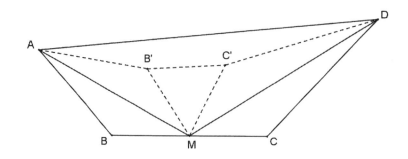

Since

$$\begin{aligned}
\angle B'MC' &= \angle AMD - \angle AMB' - \angle DMC' \\
&= 120° - (\angle AMB + \angle DMC) \\
&= 120° - (180° - 120°) \\
&= 60°,
\end{aligned}$$

$$B'M = BM = \frac{1}{2}BC = CM = C'M,$$

the triangle $B'C'M$ is equilateral, therefore $B'C' = B'M = \frac{1}{2}BC$.

Hence,

$$AD \leq AB' + B'C' + C'D = AB + \frac{1}{2}BC + CD.$$

Problem 2.5 Let O be the center of a unit square. Let O be a vertex of a bigger square. What is the area of the overlap region of the two squares?

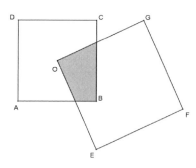

Answer

1/4

Solution

Extend the edges of the bigger square through O to divide the unit square into 4 regions, one of which is the overlap region, as shown.

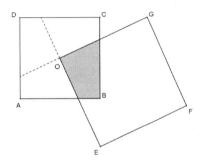

Clearly the 4 regions are all congruent. Since the unit square has area 1, the overlap region has area $1/4$.

Problem 2.6 In quadrilateral $ABCD$, $\angle ADC = \angle ABC = 90°$, $AD = CD$, $\overline{DP} \perp \overline{AB}$ at P. Assume the area of $ABCD$ is 18, find DP.

Answer

$3\sqrt{2}$

Solution

Rotate $\triangle ADP$ about point D by $90°$ to $\triangle CDP'$, as shown in the diagram.

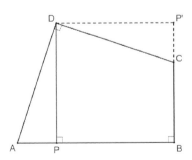

The quadrilateral $DPBP'$ is a square, with the same area as $ABCD$, which equals 18. Therefore
$$DP = \sqrt{18} = 3\sqrt{2}.$$

Problem 2.7 The side length of square $ABCD$ is 1. Let P and Q be points on $\overline{AB}$ and $\overline{AD}$ respectively, and assume the perimeter of $\triangle APQ$ is 2. Find the measure of $\angle PCQ$ in degrees.

Answer

$45°$

Solution

Rotate $\triangle DCQ$ $90°$ about point C to $\triangle BCQ'$, as shown.

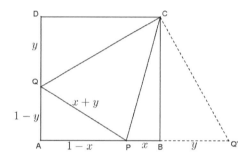

Let $BP = x$, $DQ = y$, then $AP = 1 - x$, $AQ = 1 - y$. Since the perimeter of $\triangle APQ$ is 2,

$$PQ = 2 - AP - AQ = 2 - (1 - x) - (1 - y) = x + y.$$

Also we know that $BQ' = DQ = y$, so $PQ' = x + y = PQ$.

In $\triangle CPQ$ and $\triangle CPQ'$, $\overline{CP}$ is shared, $PQ = PQ'$, and $CQ = CQ'$ because of rotation, then $\triangle CPQ \cong \triangle CPQ'$ by SSS congruency.

Since $\angle QCQ' = 90°$, we get that $\angle PCQ = \angle PCQ' = 45°$.

Problem 2.8 Given isosceles trapezoid $ABCD$, where $\overline{AD} \parallel \overline{BC}$. Construct adjacent parallelograms $ADEF$ and $CDEG$ outside the trapezoid $ABCD$. Show that $BD = FG$.

Solution

Translate FG to AC, as shown in the diagram.

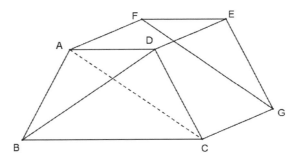

In more details: in $\triangle FEG$ and $\triangle ADC$, $FE = AD$, $EG = DC$, and $\angle FEG = \angle ADC$, so $\triangle FEG \cong \triangle ADC$, thus $FG = AC$. Also $ABCE$ is an isosceles trapezoid, so the diagonals $AC = BD$. Therefore $FG = BD$.

Problem 2.9 In $\triangle ABC$, let M be the midpoint of $\overline{BC}$. Two lines, both passing through M, and perpendicular to each other, intersect sides $\overline{AB}$ and $\overline{AC}$ at D and E respectively. Show that $BD + CE > DE$.

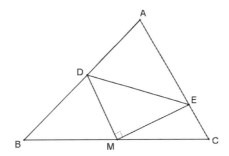

Solution

Extend $\overline{DM}$ to D' so that $D'M = DM$, connect $\overline{CD'}$ and $\overline{ED'}$, as shown.

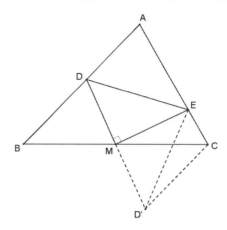

Then $CD' = BD$ and $ED' = ED$. By Triangle Inequality $CD' + CE > D'E$, and therefore $BD + CE > DE$.

Problem 2.10 Let $\triangle ABC$ be equilateral triangle, and P be a point in the interior of $\triangle ABC$. Assume that $\angle APB = 115°$, $\angle BPC = 131°$. (1) Show that AP, BP, CP can be the three sides of a triangle; (2) find the degree measures of the angles of the triangle using AP, BP, CP as sides.

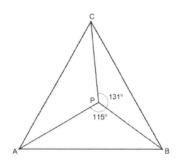

(2) 71°, 55°, and 54°

Solution

Rotate $\triangle ABP$ by 60° to $\triangle CBP'$, and connect $\overline{PP'}$, as shown.

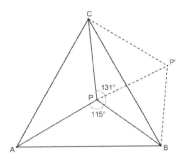

From the rotation, $CP' = AP$, and $\angle CP'B = \angle APB = 115°$.

It is easy to see that $\triangle BPP'$ is equilateral, so $PP' = BP$. Thus the triangle CPP' has the lengths of $\overline{AP}$, $\overline{BP}$, and $\overline{CP}$ as sides.

Now we compute the angles of $\triangle CPP'$:

$$\angle CPP' = \angle BPC - \angle BPP' = 131° - 60° = 71°,$$

$$\angle CP'P = \angle CP'B - \angle PP'B = 115° - 60° = 55°,$$

and

$$\angle PCP' = 180° - 71° - 55° = 54°.$$

Therefore the angles of the triangle using AP, BP, CP as sides are $71°$, $55°$, and $54°$.

Problem 2.11 In isosceles triangle ABC, $\overline{BC}$ is the base, and D is the midpoint of $\overline{BC}$. Let E be a point in the interior of $\triangle ABD$. Show that $\angle AEB > \angle AEC$.

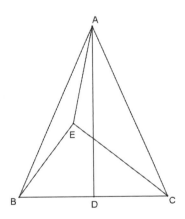

Solution

Rotate $\triangle ABE$ to $\triangle ACE'$, and connect $\overline{EE'}$, as shown.

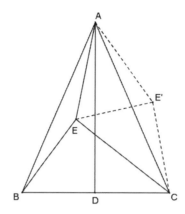

Since E is the in the interior of $\triangle ABD$, $CE > BE$. From the rotation, we know that $\angle AE'C = \angle AEB$, $AE' = AE$, and $CE' = BE < CE$. Hence, $\angle CE'E > \angle CEE'$ and $\angle AE'E = \angle AEE'$, therefore

$$\angle AEB = \angle AE'C = \angle AE'E + \angle CE'E > \angle AEE' + \angle CEE' = \angle AEC.$$

3   Solutions to Chapter 3 Examples

Problem 3.1 Let H and O be the orthocenter and circumcenter of triangle ABC. Let $\overline{ON} \perp \overline{BC}$ at N, find $\dfrac{AH}{ON}$.

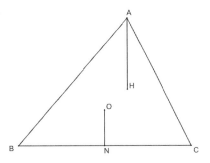

Answer

2

Solution

Construct $\overline{OM} \perp \overline{AB}$ at M, and connect $\overline{MN}$ and $\overline{CH}$, as shown.

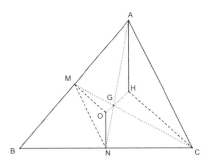

Since M and N are the midpoints of $\overline{AB}$ and $\overline{BC}$ respectively, $\overline{MN} \parallel \overline{AC}$ and $MN = \dfrac{1}{2}AC$. Also, $\overline{ON} \parallel \overline{AH}$ and $\overline{OM} \parallel \overline{CH}$, then $\triangle OMN \sim \triangle HCA$, thus

$$\frac{AH}{ON} = \frac{AC}{MN} = 2.$$

Note: Since $\overline{CM}$ and $\overline{AN}$ are medians of $\triangle ABC$, let the intersection of these medians (centroid) be G, then

$$\frac{CG}{GM} = \frac{AG}{GN} = 2.$$

It is also easy to show that $\overline{HO}$ also passes through G, and

$$\frac{HG}{GO} = 2,$$

therefore the triangles AHC and NOM are homothetic with the centroid G as the center.

Problem 3.2 In triangle ABC, from vertex A construct lines perpendicular to the angle bisectors of $\angle ABC$, $\angle ACB$ and their exterior angles, with feet D, E, F, G respectively, as shown in the diagram. Are the four points D, E, F, G collinear?

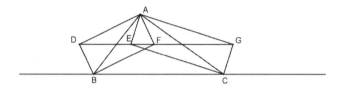

Answer

They are collinear

Solution

Extend $\overline{AD}$, $\overline{AE}$, $\overline{AF}$, and $\overline{AG}$ to intersect line $\overline{BC}$ at P. Q, R, and S respectively, as shown.

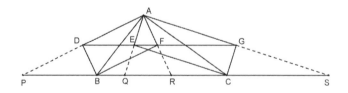

Since $\triangle BAP$, $\triangle CAQ$, $\triangle BAR$, and $\triangle CAS$ are all isosceles, we get $AD:AP = AE:AQ = AF:AR = AG:AS = 1:2$, so it is a homothety, and since P,Q,R,S are collinear, their corresponding points D, E, F, G must be too.

Problem 3.3 In acute triangle ABC, construct a rectangle $DEFG$, so that $\overline{DE}$ lies on $\overline{BC}$, and vertices G and F are on $\overline{AB}$ and $\overline{AC}$ respectively, satisfying $DE : EF = 1 : 2$.

Solution

Analysis: It is not easy to construct the desired rectangle in one shot, so we do it in two steps: first construct a $1 : 2$ rectangle $D_1E_1F_1G_1$ such that D_1 and E_1 are on $\overline{BC}$, and G_1 is on $\overline{AB}$. Then use homothety to find F on $\overline{AC}$ and the points D, E, G are found accordingly.

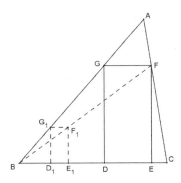

The construction is performed in the following steps. Some steps require more details to describe the movement of compass and ruler, but here we just show the results.

(1) Select a point D_1 on $\overline{BC}$ arbitrarily;

(2) Construct line $\overline{D_1G_1}$ perpendicular to $\overline{BC}$, intersecting $\overline{AC}$ at G_1;

(3) Find point E_1 on $\overline{BC}$ to the right of point D_1 such that $D_1E_1 = \dfrac{1}{2}D_1G_1$;

(4) Find point F_1 such that $D_1E_1F_1G_1$ is a rectangle (construct perpendicular line at E_1 to line $\overline{BC}$, and perpendicular line at G_1 to line G_1D_1, intersecting at F_1);

(5) Connect and extend $\overline{AF_1}$ to intersect $\overline{AC}$ at F;

(6) Construct line $\overline{FG}$ parallel to $\overline{BC}$ intersecting $\overline{AC}$ at G;

(7) Construct line $\overline{FE} \perp \overline{BC}$ at E;

(8) Construct line $\overline{GD} \perp \overline{BC}$ at D;

(9) The rectangle $DEFG$ satisfy the requirement.

The proof part of a construction problem is used to show the construction process indeed produce the desired result. The proof for this question is quite easy and is left to the reader as exercise.

Problem 3.4 Let A be a fixed point on a fixed circle $\odot O$. Point P is a point on a chord having A as one endpoint, and divides the chord into two segments with ratio $m : n$. Find the locus of P.

Answer

The locus of point P is the circle with AP' as diameter

Solution

From A construct the diameter $\overline{AB}$ of $\odot O$. Let P' be the point that divide segment $\overline{AB}$ into two parts with $m : n$ ratio, as shown.

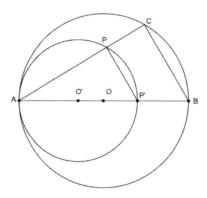

Let $\overline{AC}$ be an arbitrary chord in $\odot O$, and P be the point that divides $\overline{AC}$ into $2 : 1$ ratio. Since $\overline{AB}$ is the diameter, $\angle ACB = 90°$.
Because $AP : PC = AP' : P'B = m : n$, $\overline{PP'} \parallel \overline{CB}$, therefore $\angle APP' = 90°$. This means that the locus of point P is the circle with AP' as diameter.

Problem 3.5 As shown in the diagram, three squares are lined up. The side lengths of the two smaller squares are 3 and 2 respectively. Find the area of $\triangle ABC$.

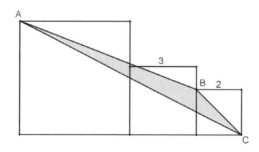

Answer

5

Solution

Draw the diagonal $\overline{AM}$ of the big square, connect $\overline{BM}$.

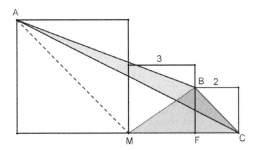

It is clear that $\overline{AM} \parallel \overline{BC}$, thus $[ABC] = [MBC]$. Since $[MBC] = \frac{1}{2}MC \cdot FB = \frac{1}{2} \times 5 \times 2 = 5$, the final answer is 5.

Problem 3.6 As shown in the diagram, in polygon $ABECD$, $\overline{AB} \parallel \overline{CD}$. Find a point C' on $\overline{CD}$ such that trapezoid $ABC'D$ has the same area as polygon $ABECD$.

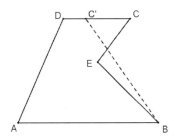

Solution

Connect $\overline{BC}$, and from point E construct a line parallel to the $\overline{BC}$, intersecting $\overline{AB}$ and $\overline{CD}$ at B' and C' respectively. Then C' is the point we need.

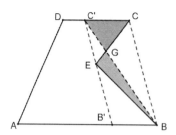

To prove that what we chose was right, note that the lines $\overline{BC}$ and $\overline{B'C'}$ are parallel, and so $[CC'E] = [BC'E]$, thus $[CC'G] = [BEG]$. Then it is clear that $[ABC'D] = [ABECD]$.

Problem 3.7 Construct a square that has the same area as a given trapezoid.

Solution

We give a sketch of the construction steps. Details of each step are left to the reader.

(1) Find the midpoints of both the legs of the trapezoid, and connect them to form the midsegment m;

(2) Find the height h of the trapezoid;

(3) Since the area of the trapezoid equals mh, it remains to find the length a such that $a^2 = mh$;

(4) On the same straight line, find three points A, B, and C in this order so that $AB = m$ and $BC = h$;

(5) Construct a semicircle using $\overline{AC}$ as diameter;

(6) From point B construct perpendicular line to $\overline{AC}$ and intersect the semicircle at D;

(7) The length CD is the value a, so construct a square with side length equal CD. This square has the same area as the given trapezoid.

Again, the proof of the correctness of this construction is left to the reader as exercise.

Problem 3.8 In trapezoid $ABCD$, $\overline{AD} \parallel \overline{BC}$, and M and N are midpoints of $\overline{AD}$ and $\overline{BC}$ respectively. Let P be an arbitrary point on $\overline{MN}$. Show that $[ABP] = [CDP]$.

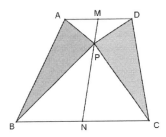

Solution

Since M is the midpoint of $\overline{AD}$, $[AMP] = [DMP]$. Similarly $[BNP] = [CNP]$.
Also the trapezoids $ABNM$ and $DCNM$ have equal bases and equal heights, so they have the same area: $[ABMN] = [DCNM]$.
Therefore,

$$[ABP] = [ABNM] - [AMP] - [BNP] = [DCNM] - [DMP] - [CNP] = [CDP].$$

4   Solutions to Chapter 4 Examples

Problem 4.1 Assume the variable x in each question is in the valid domain.

(a) $\arcsin x + \arccos x = ?$

Answer

$\pi/2$

Solution

First evaluate

$$
\begin{aligned}
&\sin(\arcsin x + \arccos x)\\
={}&\sin(\arcsin x)\cos(\arccos x) + \cos(\arcsin x)\sin(\arccos x)\\
={}&x \cdot x + \sqrt{1-x^2} \cdot \sqrt{1-x^2}\\
={}&x^2 + 1 - x^2\\
={}&1.
\end{aligned}
$$

Since $-\dfrac{\pi}{2} \le \arcsin x \le \dfrac{\pi}{2}$, and $0 \le \arccos x \le \pi$, we get

$$-\frac{\pi}{2} \le \arcsin x + \arccos x \le \frac{3\pi}{2}.$$

In the interval $\left\{ -\dfrac{\pi}{2}, \dfrac{3\pi}{2} \right\}$, the only angle whose sine value equals 1 is $\dfrac{\pi}{2}$, thus

$$\arcsin x + \arccos x = \frac{\pi}{2}.$$

(b) $\arccos x + \arccos(-x) = ?$

Answer

π

Solution

Let $\theta = \arccos x$, then $\cos\theta = x$. So $\cos(\pi - \theta) = -x$. Since the range of the function $\arccos x$ is $[0, \pi]$, and $\pi - \theta$ is in this range, $\arccos(-x) = \pi - \theta$. Therefore

$$\arccos x + \arccos(-x) = \pi.$$

Problem 4.2 Evaluate the following:

(a) $\sin 405°$

Answer

$\sqrt{2}/2$

Solution

$\sin 405° = \sin(405° - 360°) = \sin 45° = \dfrac{\sqrt{2}}{2}.$

(b) $\cos 225°$

Answer

$-\sqrt{2}/2$

Solution

$\cos 225° = \cos(180° + 45°) = -\cos 45° = -\sqrt{2}/2.$

(c) $\sin 75°$

Answer

$(\sqrt{6} + \sqrt{2})/4$

Solution

$\sin 75° = \sin(45° + 30°) = \sin 45° \cos 30° + \cos 45° \sin 30° = (\sqrt{6} + \sqrt{2})/4.$

(d) $\cot 67.5°$

Answer

$\sqrt{2} - 1$

Solution

Using the Half-Angle Formula,

$$\cot 67.5° = \frac{1}{\tan 67.5°} = \frac{1 + \cos 135°}{\sin 135°} = \frac{1 - \sqrt{2}/2}{\sqrt{2}/2} = \sqrt{2} - 1.$$

(e) $\cos\left(2\arcsin\frac{3}{5}\right)$

Answer

7/25

Solution

$$\cos\left(2\arcsin\frac{3}{5}\right) = 1 - 2\sin^2\left(\arcsin\frac{3}{5}\right) = 1 - 2\cdot\frac{9}{25} = \frac{7}{25}.$$

(f) $\sin\left(2\arccos\frac{4}{5}\right)$

Answer

24/25

Solution

Let $\theta = \arccos\frac{4}{5}$, then $\cos\theta = \frac{4}{5}$, and $\sin\theta = \frac{3}{5}$. Thus

$$\sin\left(2\arccos\frac{4}{5}\right) = \sin 2\theta = 2\sin\theta\cos\theta = 2\cdot\frac{3}{5}\cdot\frac{4}{5} = \frac{24}{25}.$$

Problem 4.3 Compute the following values.

(a) $\cos^4\frac{\pi}{24} - \sin^4\frac{\pi}{24}$

Answer

$(\sqrt{2} + \sqrt{6})/4$

Solution

$$\cos^4 \frac{\pi}{24} - \sin^4 \frac{\pi}{24} = \left(\cos^2 \frac{\pi}{24} + \sin^2 \frac{\pi}{24}\right)\left(\cos^2 \frac{\pi}{24} - \sin^2 \frac{\pi}{24}\right)$$

$$= 1 \cdot \cos \frac{\pi}{12}$$

$$= \cos\left(\frac{\pi}{3} - \frac{\pi}{4}\right)$$

$$= \cos \frac{\pi}{3} \cos \frac{\pi}{4} + \sin \frac{\pi}{3} \sin \frac{\pi}{4}$$

$$= \frac{\sqrt{2} + \sqrt{6}}{4}.$$

(b) $\cot 70° + 4\cos 70°$

Answer

$\sqrt{3}$

Solution

$$\cot 70° + 4\cos 70° = \frac{\cos 70° + 4\cos 70° \sin 70°}{\sin 70°}$$

$$= \frac{\cos 70° + 2\sin 140°}{\sin 70°}$$

$$= \frac{\cos 70° + 2\sin 40°}{\sin 70°}$$

$$= \frac{\cos 70° + 2\sin(70° - 30°)}{\sin 70°}$$

$$= \frac{\cos 70° + 2\sin 70° \cos 30° - 2\cos 70° \sin 30°}{\sin 70°}$$

$$= \frac{\cos 70° + \sin 70° \cdot \sqrt{3} - \cos 70°}{\sin 70°}$$

$$= \frac{\sqrt{3}\sin 70°}{\sin 70°}$$

$$= \sqrt{3}.$$

(c) $\sin 10° \sin 30° \sin 50° \sin 70°$

Answer

1/16

Solution

$$
\begin{aligned}
\sin 10° \sin 30° \sin 50° \sin 70° &= \frac{1}{2} \sin 10° \sin 50° \sin 70° \\
&= \frac{1}{2} \cos 20° \cos 40° \cos 80° \\
&= \frac{\sin 20° \cos 20° \cos 40° \cos 80°}{2 \sin 20°} \\
&= \frac{\sin 40° \cos 40° \cos 80°}{4 \sin 20°} \\
&= \frac{\sin 80° \cos 80°}{8 \sin 20°} \\
&= \frac{\sin 160°}{16 \sin 20°} \\
&= \frac{\sin 20°}{16 \sin 20°} \\
&= \frac{1}{16}.
\end{aligned}
$$

(d) $\arctan \dfrac{1}{2} + \arctan \dfrac{1}{3}$

Answer

$\pi/4$ or $45°$

Solution

Let $\alpha = \arctan\dfrac{1}{2}$, $\beta = \arctan\dfrac{1}{3}$, use the sum formula for $\tan(\alpha + \beta)$:

$$
\begin{aligned}
\tan(\alpha + \beta) &= \frac{\tan\alpha + \tan\beta}{1 - \tan\alpha\tan\beta} \\
&= \frac{\dfrac{1}{2} + \dfrac{1}{3}}{1 - \dfrac{1}{2}\cdot\dfrac{1}{3}} \\
&= 1,
\end{aligned}
$$

thus $\alpha + \beta = \dfrac{\pi}{4}$.

(e) $(1 - \cot 1^\circ)(1 - \cot 44^\circ)$

Answer

2

Solution

Use the following identity:

$$
\cot(\alpha + \beta) = \frac{\cot\alpha + \cot\beta}{\cot\alpha\cot\beta - 1}.
$$

Thus

$$
1 = \cot 45^\circ = \cot(1^\circ + 44^\circ) = \frac{\cot 1^\circ + \cot 44^\circ}{\cot 1^\circ \cot 44^\circ - 1},
$$

so

$$
\cot 1^\circ \cot 44^\circ = \cot 1^\circ + \cot 44^\circ + 1,
$$

therefore

$$
\begin{aligned}
(1 - \cot 1^\circ)(1 - \cot 44^\circ) &= 1 - \cot 1^\circ - \cot 44^\circ + \cot 1^\circ \cot 44^\circ \\
&= 1 - \cot 1^\circ - \cot 44^\circ + \cot 1^\circ + \cot 44^\circ + 1 \\
&= 2.
\end{aligned}
$$

Problem 4.4 The quadratic equation $2x^2 - (\sqrt{3}+1)x + m = 0$ has two roots $\sin\theta$ and $\cos\theta$, find the value of $\dfrac{\sin\theta}{1-\cot\theta} + \dfrac{\cos\theta}{1-\tan\theta}$.

Answer

$\dfrac{\sqrt{3}+1}{2}$

Solution

By Vieta's formulas, $\sin\theta + \cos\theta = \dfrac{\sqrt{3}+1}{2}$, thus

$$
\begin{aligned}
\frac{\sin\theta}{1-\cot\theta} + \frac{\cos\theta}{1-\tan\theta} &= \frac{\sin^2\theta}{\sin\theta - \cos\theta} + \frac{\cos^2\theta}{\cos\theta - \sin\theta} \\
&= \frac{\sin^2\theta - \cos^2\theta}{\sin\theta - \cos\theta} \\
&= \sin\theta + \cos\theta \\
&= \frac{\sqrt{3}+1}{2}.
\end{aligned}
$$

Problem 4.5 Given an equilateral triangle and its incircle. An arc on the incircle has the same length as the side length of the equilateral triangle. What is the measure, in radians, of the arc?

Answer

$2\sqrt{3}$

Solution

Let r be the inradius of the equilateral triangle, then the side length of the equilateral triangle is $2\sqrt{3}r$. Thus the arc has length $2\sqrt{3}r$, and then its angular measure is $2\sqrt{3}$ radians.

Problem 4.6 Given that α and β are acute angles, satisfying

$$
\cos\alpha + \cos\beta - \cos(\alpha+\beta) = \frac{3}{2},
$$

find the sum of all possible values of $\alpha + \beta$ in degrees. (ZIML Varsity March 2018)

Answer

120

Solution

Use the sum-to-product formula and double-angle formula,

$$2\cos\frac{\alpha+\beta}{2}\cos\frac{\alpha-\beta}{2} - 2\cos^2\frac{\alpha+\beta}{2} + 1 = \frac{3}{2},$$

so

$$4\cos\frac{\alpha+\beta}{2}\cos\frac{\alpha-\beta}{2} - 4\cos^2\frac{\alpha+\beta}{2} - 1 = 0,$$

Change signs for all terms, and complete the square,

$$\left(2\cos\frac{\alpha+\beta}{2} - \cos\frac{\alpha-\beta}{2}\right)^2 + \sin^2\frac{\alpha-\beta}{2} = 0.$$

Thus

$$2\cos\frac{\alpha+\beta}{2} - \cos\frac{\alpha-\beta}{2} = 0,$$

and

$$\sin\frac{\alpha-\beta}{2} = 0.$$

Since α and β are both acute, $-90° < \dfrac{\alpha-\beta}{2} < 90°$, thus $\alpha - \beta = 0$, that is, $\alpha = \beta$, and $\cos\alpha = \dfrac{1}{2}$, thus $\alpha = 60°$. Also we get $\beta = 60°$ as well. Therefore, $\alpha + \beta = 120°$.

Problem 4.7 Let $0 < \theta < \pi$, find the maximum value of $\sin\dfrac{\theta}{2}(1+\cos\theta)$. (Do not use calculus methods)

Answer

$\dfrac{4\sqrt{3}}{9}$

Solution

We apply the AM-GM inequality, using the fact that $\sin\dfrac{\theta}{2} > 0$ in the interval $0 < \theta < \pi$. Let

$$A = \sin\frac{\theta}{2}(1+\cos\theta) = 2\sin\frac{\theta}{2}\cos^2\frac{\theta}{2},$$

then

$$A^2 = 4\sin^2\frac{\theta}{2}\cos^4\frac{\theta}{2}$$

$$= 16\left(\sin^2\frac{\theta}{2}\right)\cdot\left(\frac{1}{2}\cos^2\frac{\theta}{2}\right)\cdot\left(\frac{1}{2}\cos^2\frac{\theta}{2}\right)$$

$$\le 16\left(\frac{\sin^2\frac{\theta}{2}+\frac{1}{2}\cos^2\frac{\theta}{2}+\frac{1}{2}\cos^2\frac{\theta}{2}}{3}\right)^3$$

$$= 16\left(\frac{1}{3}\right)^3$$

$$= \frac{16}{27},$$

therefore

$$A \le \sqrt{\frac{16}{27}} = \frac{4\sqrt{3}}{9}.$$

Note: equality occurs when $\sin^2\frac{\theta}{2}=\frac{1}{2}\cos^2\frac{\theta}{2}$, i.e. $\theta = 2\arctan\frac{\sqrt{2}}{2}$.

Problem 4.8 In $\triangle ABC$, let a,b,c be the lengths of the sides opposite angles A,B,C respectively. If $c-a$ equals the altitude h on side $\overline{AC}$, find the value of $\sin\frac{C-A}{2}+\cos\frac{C+A}{2}$.

Answer

1

Solution

Since h is the altitude on side $\overline{AC}$, $h = a\sin C = c\sin A$, therefore $a = \dfrac{h}{\sin C}$ and $c = \dfrac{h}{\sin A}$. Based on the question,

$$h = c - a = \frac{h}{\sin A} - \frac{h}{\sin C},$$

we get that

$$1 = \frac{1}{\sin A} - \frac{1}{\sin C},$$

so
$$\sin A \sin C = \sin C - \sin A.$$

Applying the product-to-sum and sum-to-product formulas,
$$-\frac{1}{2}(\cos(C+A) - \cos(C-A)) = 2\cos\frac{C+A}{2}\sin\frac{C-A}{2},$$

From Double-Angle formula,
$$\cos(C+A) - \cos(C-A) = \left(2\cos^2\frac{C+A}{2} - 1\right) - \left(1 - 2\sin^2\frac{C-A}{2}\right)$$
$$= 2\left(\cos^2\frac{C+A}{2} + \sin^2\frac{C-A}{2} - 1\right),$$

thus
$$1 - \sin^2\frac{C+A}{2} - \cos^2\frac{C-A}{2} = 2\cos\frac{C+A}{2}\sin\frac{C-A}{2},$$

so
$$\left(\sin\frac{C-A}{2} + \cos\frac{C+A}{2}\right)^2 - 1 = 0,$$

and then
$$\left(\sin\frac{C-A}{2} + \cos\frac{C+A}{2} + 1\right)\left(\sin\frac{C-A}{2} + \cos\frac{C+A}{2} - 1\right) = 0.$$

Since $\sin\dfrac{C-A}{2} > 0$, and $\cos\dfrac{C+A}{2} > 0$
$$\sin\frac{C-A}{2} + \cos\frac{C+A}{2} + 1 > 0,$$

therefore
$$\sin\frac{C-A}{2} + \cos\frac{C+A}{2} - 1 = 0,$$

that is
$$\sin\frac{C-A}{2} + \cos\frac{C+A}{2} = 1.$$

Problem 4.9 Given $\alpha \in \left(\dfrac{\pi}{4}, \dfrac{\pi}{2}\right)$, list the following in increasing order:
$$(\cos\alpha)^{\cos\alpha}, (\sin\alpha)^{\cos\alpha}, (\cos\alpha)^{\sin\alpha}.$$

Answer

$(\cos\alpha)^{\sin\alpha} < (\cos\alpha)^{\cos\alpha} < (\sin\alpha)^{\cos\alpha}$

Solution

Since $\cos\alpha < 1$, the exponential function $f(x) = (\cos\alpha)^x$ is a decreasing function for all real numbers x. For $\alpha \in \left(\dfrac{\pi}{4}, \dfrac{\pi}{2}\right)$, $\sin\alpha > \cos\alpha$, so

$$(\cos\alpha)^{\sin\alpha} < (\cos\alpha)^{\cos\alpha}.$$

On the other hand, $\cos\alpha > 0$, so the power function $g(x) = x^{\cos\alpha}$ is an increasing function for $x > 0$, thus

$$(\cos\alpha)^{\cos\alpha} < (\sin\alpha)^{\cos\alpha}.$$

Combining the conclusions,

$$(\cos\alpha)^{\sin\alpha} < (\cos\alpha)^{\cos\alpha} < (\sin\alpha)^{\cos\alpha}.$$

Problem 4.10 Let α and β be acute angles and $\alpha + \beta = 90°$, also $\sin\alpha$ and $\sin\beta$ are the roots of the equation $2x^2 - 2\sqrt{2}x + c = 0$. Suppose α equals k degrees, What is the value of $c + k$? (ZIML Varsity March 2018)

Answer

46

Solution

By Vieta's formulas, $\sin\alpha + \sin\beta = \sqrt{2}$ and $\sin\alpha \cdot \sin\beta = \dfrac{c}{2}$. Therefore,

$$
\begin{aligned}
(\sin\alpha + \sin\beta)^2 &= 2, \\
\sin^2\alpha + 2\sin\alpha\sin\beta + \sin^2\beta &= 2, \\
1 + c &= 2, \\
c &= 1.
\end{aligned}
$$

Also since $\alpha + \beta = 90°$, $\sin\beta = \cos\alpha$, so

$$\sqrt{2} = \sin\alpha + \cos\alpha = \sqrt{2}\sin(\alpha + 45°),$$

thus $\sin(\alpha + 45°) = 1$. Based on the fact that α is an acute angle, $\alpha + 45° = 90°$, then $\alpha = 45°$. Hence, $c = 1$ and $k = 45$, and the final answer is 46.

Problem 4.11 Evaluate

$$\sin^2 80° + \sin^2 40° - \cos 50° \cos 10°,$$

express your answer in decimal, rounded to the nearest hundredth if necessary. (ZIML Varsity June 2018)

Answer

0.75

Solution

Complete the square, and apply the Sum-to-Product and Product-to-Sum formulas,

$$\begin{aligned}
\sin^2 80° + \sin^2 40° - \cos 50° \cos 10° &= (\sin 80° - \sin 40°)^2 + \cos 50° \cos 10° \\
&= 2^2 \sin^2 20° \cos^2 60° + \frac{1}{2}(\cos 60° + \cos 40°) \\
&= \sin^2 20° + \frac{1}{4} + \frac{1}{2}\cos 40° \\
&= \frac{1}{2}(1 - \cos 40°) + \frac{1}{4} + \frac{1}{2}\cos 40° \\
&= \frac{1}{2} + \frac{1}{4} \\
&= 0.75.
\end{aligned}$$

5   Solutions to Chapter 5 Examples

Problem 5.1 In $\triangle ABC$, let D and E be points on $\overline{BC}$ and $\overline{AB}$ respectively, such that $\dfrac{BD}{DC} = \dfrac{1}{3}$ and $\dfrac{AE}{EB} = \dfrac{2}{3}$. Also let G be on $\overline{AD}$ such that $\dfrac{AG}{GD} = \dfrac{1}{2}$. Let F be the intersection of $\overline{EG}$ and $\overline{AC}$. Find $\dfrac{AF}{FC}$.

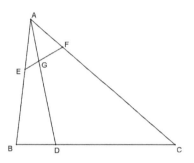

Answer

$2/7$

Solution

Extend $\overline{CB}$ and $\overline{FE}$ to intersect at H.

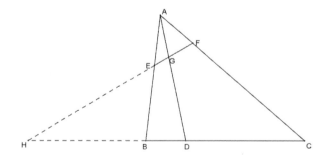

Line $\overline{GEH}$ intersects the three sides of $\triangle ADB$, so by Menelaus' Theorem,

$$\frac{AG}{GD} \cdot \frac{DH}{HB} \cdot \frac{BE}{EA} = \frac{1}{2} \cdot \frac{DH}{HB} \cdot \frac{3}{2} = 1,$$

thus $\dfrac{DH}{HB} = \dfrac{4}{3}$, and then since $\dfrac{BD}{DC} = \dfrac{1}{3}$, $\dfrac{CH}{HD} = \dfrac{7}{4}$.

Line $\overline{FGH}$ intersects the three sides of $\triangle ACD$, and by Menelaus' Theorem,

$$\frac{AF}{FC} \cdot \frac{CH}{HD} \cdot \frac{DG}{GA} = \frac{AF}{FC} \cdot \frac{7}{4} \cdot \frac{2}{1} = 1,$$

Therefore $\dfrac{AF}{FC} = \dfrac{2}{7}$.

Problem 5.2 In parallelogram $ABCD$, let E and F be the midpoints of $\overline{AB}$ and $\overline{BC}$ respectively. $\overline{AF}$ and $\overline{CE}$ intersect at G, and $\overline{AF}$ and $\overline{DE}$ intersect at H. Find the ratio $AH : HG : GF$.

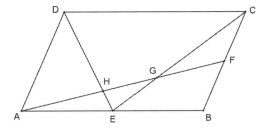

Answer

$6 : 4 : 5$

Solution

Extend $\overline{CB}$ and $\overline{DE}$ and intersect at P.

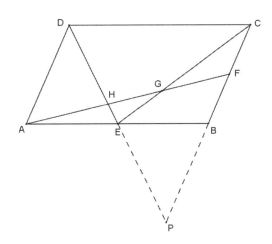

It is easy to see that $BP = BC$, and so $\dfrac{FP}{PB} = \dfrac{3}{2}$. For $\triangle AFB$ and line $\overline{HEP}$, apply Menelaus's Theorem, so

$$\frac{AH}{HF} \cdot \frac{FP}{PB} \cdot \frac{BE}{EA} = 1,$$

thus $\dfrac{AH}{HF} = \dfrac{2}{3}$, and then $\dfrac{AH}{AF} = \dfrac{2}{5}$.

Similarly, apply Menelaus Theorem on $\triangle AFB$ and line $\overline{CGE}$,

$$\frac{AG}{GF} \cdot \frac{FC}{CB} \cdot \frac{BE}{EA} = 1,$$

we know that $\dfrac{FC}{CB} = \dfrac{1}{2}$, thus $\dfrac{AG}{GF} = 2$, so $\dfrac{AG}{AF} = \dfrac{2}{3}$.

Now we can calculate

$$\frac{HG}{AF} = \frac{AG}{AF} - \frac{AH}{AF} = \frac{2}{3} - \frac{2}{5} = \frac{4}{15},$$

and

$$\frac{GF}{AF} = 1 - \frac{AG}{AF} = \frac{1}{3}.$$

Therefore

$$AH : HG : GF = \frac{2}{5} : \frac{4}{15} : \frac{1}{3} = 6 : 4 : 5.$$

Problem 5.3 (AIME 1989) Let P be a point in the interior of $\triangle ABC$. Connect and extend $\overline{AP}, \overline{BP}, \overline{CP}$ and intersect $\overline{BC}, \overline{CA}, \overline{AB}$ at D, E, F respectively. Given that $AP = 6, BP = 9, PD = 6, PE = 3, CF = 20$, find the area of $\triangle ABC$.

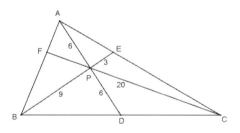

Answer

108

Solution

Since line $\overline{CEA}$ intersects the extensions of the sides of $\triangle BDP$, by Menelaus' Theorem,

$$\frac{BC}{CD} \cdot \frac{DA}{AP} \cdot \frac{PE}{EA} = \frac{BC}{CD} \cdot \frac{12}{6} \cdot \frac{3}{12} = 1,$$

therefore $\frac{BC}{CD} = 2$, which means $BD = DC$.

Similarly, line $\overline{BFA}$ intersects the extensions of the sides of $\triangle CDP$, by Menelaus' Theorem,

$$\frac{CB}{BD} \cdot \frac{DA}{AP} \cdot \frac{PF}{FC} = \frac{2}{1} \cdot \frac{2}{1} \cdot \frac{PF}{FC} = 1,$$

therefore $\frac{PF}{FC} = \frac{1}{4}$. Since $CF = 20$, we get $PF = 5$, so $CP = 15$.

Since $\overline{PD}$ is the median of $\triangle PBC$, by the formula for the length of medians,

$$PD = \frac{1}{2}\sqrt{2(PB^2 + PC^2) - BC^2},$$

which is

$$6 = \frac{1}{2}\sqrt{2(9^2 + 15^2) - BC^2},$$

and solve to get $BC = 6\sqrt{13}$, hence $BD = 3\sqrt{13}$.
Also, since $BP^2 + PD^2 = 6^2 + 9^2 = 117 = BD^2$, by Pythagorean Theorem $\angle BPD = 90°$.
So $[BPD] = 6 \times 9/2 = 27$, and $[ABC] = 2[ABD] = 4[BPD] = 108$.

Problem 5.4 As shown in the diagram, in quadrilateral $ABCD$, the area ratio of $\triangle ABD, \triangle BCD$, and $\triangle ABC$ is $3:4:1$. Let M and N be points on $\overline{AC}$ and $\overline{CD}$ respectively, satisfying $\frac{AM}{AC} = \frac{CN}{CD} = r$, and B, M, N are collinear. Find the value of r.

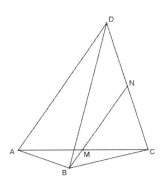

Answer

1/2

Solution

Let E be the intersection of $\overline{AC}$ and $\overline{BD}$.
Since B, M, N are collinear, apply Menelaus' Theorem on triangle DEC and line $\overline{BMN}$,

$$\frac{CN}{ND} \cdot \frac{DB}{BE} \cdot \frac{EM}{MC} = 1.$$

Now we find out the ratios in the above equation in terms of r.
Since $[ABD] : [BCD] = 3 : 4$, $AE : EC = 3 : 4$, thus $\dfrac{AE}{AC} = \dfrac{3}{7}$.

Also $[ACD] : [ABC] = (3 + 4 - 1) : 1 = 6 : 1$, thus $DE : EB = 6 : 1$, and then $\dfrac{DB}{BE} = \dfrac{7}{1}$.

To find $\dfrac{EM}{MC}$, note that $EM = AM - AE$ and $MC = AC - AM$, so

$$\frac{EM}{MC} = \frac{AM - AE}{AC - AM} = \frac{\dfrac{AM}{AC} - \dfrac{AE}{AC}}{1 - \dfrac{AM}{AC}} = \frac{r - \dfrac{3}{7}}{1 - r} = \frac{7r - 3}{7 - 7r}.$$

To find $\dfrac{CN}{ND}$, note that $\dfrac{CN}{CD} = r$, so

$$\frac{CN}{ND} = \frac{CN}{CD - CN} = \frac{r}{1 - r}.$$

Hence

$$\frac{r}{1 - r} \cdot \frac{7}{1} \cdot \frac{7r - 3}{7 - 7r} = 1,$$

simplify and get

$$6r^2 - r - 1 = 0,$$

so $r = \dfrac{1}{2}$ or $-\dfrac{1}{3}$, where $-\dfrac{1}{3}$ is extraneous.

Therefore the final answer is $\dfrac{1}{2}$.

Problem 5.5 In quadrilateral $ABCD$, the diagonal $\overline{AC}$ bisects $\angle BAD$. Let E be a point on $\overline{CD}$, connect $\overline{BE}$ and intersect $\overline{AC}$ at F, and connect and extend $\overline{DF}$ to intersect $\overline{BC}$ at G. Show that $\angle GAC = \angle EAC$.

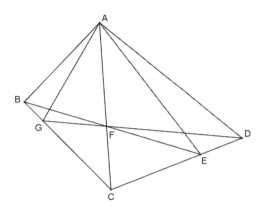

Solution 1

Let $\angle BAC = \angle CAD = \theta, \angle GAC = \alpha, \angle EAC = \beta$. We want to show that $\alpha = \beta$.
Line $\overline{GFD}$ intersects with the sides of $\triangle BCE$, so we apply Menelaus' Theorem, then

$$
\begin{aligned}
1 &= \frac{BG}{GC} \cdot \frac{CD}{DE} \cdot \frac{EF}{FB} \\
&= \frac{[ABG]}{[ACG]} \cdot \frac{[ACD]}{[AED]} \cdot \frac{[AEF]}{[ABF]} \\
&= \frac{AB\sin(\theta - \alpha)}{AC\sin\alpha} \cdot \frac{AC\sin\theta}{AE\sin(\theta - \beta)} \cdot \frac{AE\sin\beta}{AB\sin\theta} \\
&= \frac{\sin(\theta - \alpha)\sin\beta}{\sin\alpha\sin(\theta - \beta)},
\end{aligned}
$$

thus

$$
\begin{aligned}
\sin(\theta - \alpha)\sin\beta &= \sin\alpha\sin(\theta - \beta), \\
\sin\theta\cos\alpha\sin\beta - \cos\theta\sin\alpha\sin\beta &= \sin\theta\cos\beta\sin\alpha - \cos\theta\sin\beta\sin\alpha, \\
\sin\theta\cos\alpha\sin\beta &= \sin\theta\cos\beta\sin\alpha, \\
\cos\alpha\sin\beta - \cos\beta\sin\alpha &= 0, \\
\sin(\beta - \alpha) &= 0,
\end{aligned}
$$

where the only possibility is $\beta - \alpha = 0$, therefore $\angle GAC = \angle EAC$.

Solution 2

We apply the converse of Menelaus' Theorem, that is, the "if" part of Theorem 5.

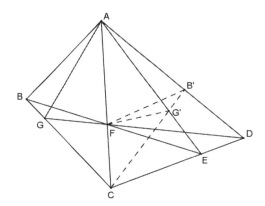

As shown in the diagram, let B' be the point on $\overline{AD}$ such that $AB' = AB$ (i.e. reflect B with respect to $\overline{AC}$ to B'). Also reflect G with respect to $\overline{AC}$ to G', so $\angle G'AC = \angle GAC$. By symmetry B', G', C are collinear. We want to show that A, G', E are collinear. Connect $\overline{FB'}$ and $\overline{FG'}$. Let $\angle EFB' = \alpha$, $\angle DFE = \angle BFG = \angle B'FG' = \beta$, and $\angle AFD = \angle GFC = \angle G'FC = \gamma$. Then consider $\triangle CDB'$ and points A, G', E,

$$
\begin{aligned}
\frac{DA}{AB'} \cdot \frac{B'G'}{G'C} \cdot \frac{CE}{ED} &= \frac{[FDA]}{[FB'A]} \cdot \frac{[FG'B']}{[FG'C]} \cdot \frac{[FEC]}{[FED]} \\[2mm]
&= \frac{FD\sin\gamma}{FB'\sin(\beta+\gamma-\alpha)} \cdot \frac{FB'\sin\beta}{FC\sin\gamma} \cdot \frac{FC\sin(\beta+\gamma-\alpha)}{FD\sin\beta} \\[2mm]
&= 1.
\end{aligned}
$$

Therefore, A, G', E are collinear, and thus $\angle GAC = \angle EAC$.

Solution 3

Connect $\overline{BD}$ and intersect $\overline{AC}$ at H.

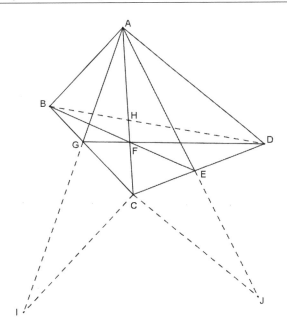

Apply Ceva's Theorem on $\triangle BCD$ and point F,

$$\frac{CG}{GB} \cdot \frac{BH}{HD} \cdot \frac{DE}{EC} = 1.$$

Since $\overline{AH}$ bisects $\angle BAD$, by Angle Bisector Theorem,

$$\frac{BH}{HD} = \frac{AB}{AD},$$

therefore

$$\frac{CG}{GB} \cdot \frac{AB}{AD} \cdot \frac{DE}{EC} = 1.$$

Through point C construct a line parallel to $\overline{AB}$, intersecting the extension of $\overline{AG}$ at I; also through C construct a line parallel to $\overline{AD}$, intersecting the extension of $\overline{AE}$ at J, then

$$\frac{CG}{GB} = \frac{CI}{AB}, \quad \frac{DE}{EC} = \frac{AD}{CJ},$$

hence

$$\frac{CI}{AB} \cdot \frac{AB}{AD} \cdot \frac{AD}{CJ} = 1,$$

which means $CI = CJ$.
Also we have $\overline{CI} \parallel \overline{AB}$, and $\overline{CJ} \parallel \overline{AD}$, so

$$\angle ACI = 180^\circ - \angle BAC = 180^\circ - \angle DAC = \angle ACJ,$$

therefore $\triangle ACI \cong \triangle ACJ$, and then $\angle IAC = \angle JAC$, which is the same as

$$\angle GAC = \angle EAC.$$

Problem 5.6 In $\triangle ABC$, D is a point on $\overline{BC}$ such that $\dfrac{BD}{DC} = \dfrac{1}{3}$. Let E be the midpoint of $\overline{AC}$, O be the intersection of $\overline{AD}$ and $\overline{BE}$, F be the intersection of $\overline{CO}$ and $\overline{AB}$. Find the ratio between the area of quadrilateral $BDOF$ and the area of $\triangle ABC$.

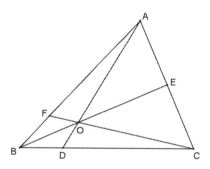

Answer

$1/10$

Solution

In $\triangle ABC$, $\overline{AD}$, $\overline{BE}$, and $\overline{CF}$ are concurrent at O, so by Ceva's Theorem,

$$\frac{BD}{DC} \cdot \frac{CE}{EA} \cdot \frac{AF}{FB} = 1,$$

Since $\dfrac{BD}{DC} = \dfrac{1}{3}$, $\dfrac{CE}{EA} = 1$, we get $\dfrac{AF}{FB} = 3$. Thus $\dfrac{AF}{AB} = \dfrac{3}{4}$.

Considering line $\overline{BOE}$ and $\triangle ADC$, we apply Menelaus' Theorem, then

$$\frac{DB}{BC} \cdot \frac{CE}{EA} \cdot \frac{AO}{OD} = 1;$$

also we know that $\dfrac{DB}{BC} = \dfrac{1}{4}$, thus $\dfrac{AO}{OD} = 4$. Hence, $\dfrac{AO}{AD} = \dfrac{4}{5}$, and consequently

$$\frac{[AFO]}{[ABD]} = \frac{AF \cdot AO \cdot \sin \angle BAD}{AB \cdot AD \cdot \sin \angle BAD} = \frac{3}{4} \cdot \frac{4}{5} = \frac{3}{5},$$

and therefore $\dfrac{[BDOF]}{[ABD]} = \dfrac{2}{5}$. Also, $\dfrac{[ABD]}{[ABC]} = \dfrac{1}{4}$, thus $\dfrac{[BDOF]}{[ABC]} = \dfrac{2}{5} \cdot \dfrac{1}{4} = \dfrac{1}{10}$.

Problem 5.7 In $\triangle ABC$, $\angle ABC = \angle ACB = 40°$. Let P be an interior point in $\triangle ABC$, such that $\angle PAC = 20°$, $\angle PCB = 30°$. Find the degree measure of $\angle PBC$.

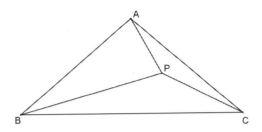

Answer

$20°$

Solution

Let $x = \angle PBC$, then $\angle ABP = 40° - x$. By the trigonometric form of Ceva's Theorem,

$$\frac{\sin 80°}{\sin 20°} \cdot \frac{\sin x}{\sin(40° - x)} \cdot \frac{\sin 10°}{\sin 30°} = 1.$$

Since $\sin 80° \sin 10° = \cos 10° \sin 10° = \dfrac{1}{2}\sin 20°$, and $\sin 30° = \dfrac{1}{2}$, the above equation becomes

$$\frac{\sin x}{\sin(40° - x)} = 1.$$

This means $\sin x = \sin(40° - x)$, so either $x = 40° - x$ which implies $x = 20°$, or $180° - x = 40° - x$ which is impossible. Therefore the final answer is $20°$.

Problem 5.8 In $\triangle ABC$, $AB = AC$, $\angle A = 80°$. Let D be a point in the interior of $\triangle ABC$ such that $\angle DAB = \angle DBA = 10°$. Find the degree measure of $\angle ACD$.

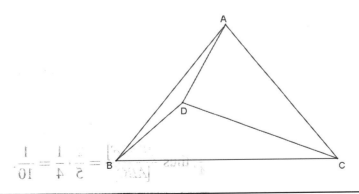

Answer

$30°$

Solution

Let $x = \angle ACD$, then $\angle BCD = 50° - x$. By the trigonometric form of Ceva's Theorem,

$$\frac{\sin 10°}{\sin 70°} \cdot \frac{\sin 40°}{\sin 10°} \cdot \frac{\sin x}{\sin(50° - x)} = 1.$$

To solve the above equation, note that $\sin 70° = \cos 20°$, and $\sin 40° = 2\sin 20° \cos 20°$, the equation becomes

$$\frac{2\sin 20° \sin x}{\sin(50° - x)} = 1,$$

which means

$$2\sin 20° \sin x = \sin(50° - x).$$

Seeing $20°$ as $50° - 30°$,

$$\begin{aligned}
2\sin(50° - 30°)\sin x &= \sin(50° - x), \\
(\sqrt{3}\sin 50° - \cos 50°)\sin x &= \sin 50° \cos x - \cos 50° \sin x, \\
\sqrt{3}\sin 50° \sin x &= \sin 50° \cos x, \\
\tan x &= \frac{1}{\sqrt{3}},
\end{aligned}$$

thus $x = 30°$.

Problem 5.9 In $\triangle ABC$, $\angle BAC = 30°$, $\angle ABC = 70°$. Let M be a point in the interior of $\triangle ABC$ such that $\angle MAB = \angle MCA = 20°$. Find the degree measure of $\angle MBA$.

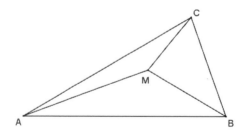

Answer

$30°$

Solution

Let $x = \angle MBA$, then $\angle MBC = 70° - x$. Also,

$$\angle MAC = \angle BAC - \angle MAB = 30° - 20° = 10°, \quad \angle ACB = 180° - \angle BAC - \angle ABC = 80°,$$

and

$$\angle MCB = \angle ACB - \angle MCA = 80° - 20° = 60°.$$

By the trigonometric form of Ceva's Theorem,

$$\frac{\sin 20°}{\sin 10°} \cdot \frac{\sin 20°}{\sin 60°} \cdot \frac{\sin(70° - x)}{\sin x} = 1.$$

Knowing that $\dfrac{\sin 20°}{\sin 10°} = \dfrac{2\sin 10° \cos 10°}{\sin 10°} = 2\cos 10° = 2\sin 80°$, the equation becomes

$$\frac{2\sin 80° \sin 20° \sin(70° - x)}{\sin 60° \sin x} = 1.$$

Now we use the identity that

$$4\sin 20° \sin 40° \sin 80° = \sin 60°.$$

(To prove this identity, apply the Product-to-Sum formulas,

$$\begin{aligned}
4\sin 20° \sin 40° \sin 80° &= 2(\cos 20° - \cos 60°)\sin 80° \\
&= 2\cos 20° \sin 80° - \sin 80° \\
&= \sin 100° + \sin 60° - \sin 80° \\
&= \sin 80° + \sin 60° - \sin 80° \\
&= \sin 60°.
\end{aligned}$$

)
So we get

$$\sin 60° \sin(70° - x) = 2\sin 40° \sin 60° \sin x.$$

Canceling $\sin 60°$, and noting that $40° = 70° - 30°$,

$$\begin{aligned}
\sin(70° - x) &= 2\sin(70° - 30°)\sin x, \\
\sin 70° \cos x - \cos 70° \sin x &= 2\sin 70° \cos 30° \sin x - 2\cos 70° \sin 30° \sin x, \\
\sin 70° \cos x - \cos 70° \sin x &= \sqrt{3}\sin 70° \sin x - \cos 70° \sin x, \\
\sin 70° \cos x &= \sqrt{3}\sin 70° \sin x, \\
\tan x &= \frac{1}{\sqrt{3}}.
\end{aligned}$$

Therefore $x = 30°$.

Problem 5.10 Let $\overline{AB}$ be a diameter of $\odot O$. Chord $\overline{CD}$ is perpendicular to $\overline{AB}$ at L, and points M and N are on segments $\overline{LB}$ and $\overline{LA}$ respectively, satisfying $LM : MB = LN : NA$. Rays $\overrightarrow{CM}$ and $\overrightarrow{CN}$ intersect $\odot O$ at E and F respectively. Show that $\overline{AE}$, $\overline{BF}$, and $\overline{OD}$ are concurrent.

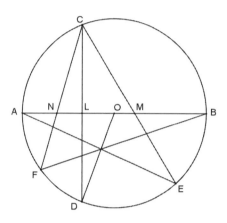

Solution

Connect $\overline{AD}$ intersecting $\overline{BF}$ at G, and connect $\overline{BD}$ intersecting $\overline{AE}$ at H. Connect $\overline{GN}$ and $\overline{HM}$, as shown in the diagram.

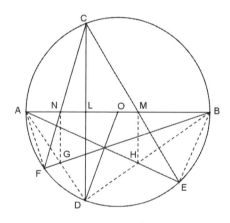

Since $\angle DAB = \angle CDB = \angle CFB$, the points A, F, G, N are concyclic, therefore $\angle AFB + \angle ANG = 180°$. Also since $\overline{AB}$ is the diameter of $\odot O$, $\angle AFB = 90°$, so $\angle ANG = 90°$, that is, $\overline{GN} \perp \overline{AB}$ and then $\overline{GN} \parallel \overline{DL}$, and thus $\dfrac{LN}{NA} = \dfrac{DG}{GA}$. Similarly, $\dfrac{HD}{BH} = \dfrac{LM}{MB}$. Given that $\dfrac{LM}{MB} = \dfrac{LN}{NA}$, we get $\dfrac{DG}{GA} = \dfrac{HD}{BH}$.

Hence in $\triangle ABD$,

$$\frac{AO}{OB} \cdot \frac{BH}{HD} \cdot \frac{DG}{GA} = 1,$$

so by Ceva's Theorem, $\overline{AE}, \overline{BF}$, and $\overline{OD}$ are concurrent.

Made in the USA
Middletown, DE
27 December 2018